D0561349

Footsteps in the Jungle

BY JONATHAN MASLOW

Books
Torrid Zone
Sacred Horses
Bird of Life, Bird of Death
The Owl Papers

Documentary Films
Saddle the Wind
A Tramp in the Darien

JONATHAN MASLOW

Footsteps in the Jungle

ADVENTURES IN THE SCIENTIFIC EXPLORATION OF THE AMERICAN TROPICS

CHICAGO *Ivan R. Dee* 1996

 For information, address: Ivan R. Dee, Inc., 1332 North Halsted Street, Chicago 60622. Manufactured in the United States of America and printed on acid-free paper.

Maps by Victor Thompson

Library of Congress Cataloging-in-Publication Data:
Maslow, Jonathan Evan.
Footsteps in the jungle : adventures in the scientific exploration of the American tropics / Jonathan Maslow.
p. cm.
Includes bibliographical references and index.
ISBN 1-56663-137-8 (alk. paper)
1. Naturalists—Latin America—Biography. 2. Explorers—Latin America—Biography. 3. Natural history—Latin America. I. Title.
QH26.M27 1996
508.8'092'2—dc20
[B] 96-19161

For Marion Zunz,

zoologist and filmmaker,

in memoriam

Contents

TERRITORIES OF

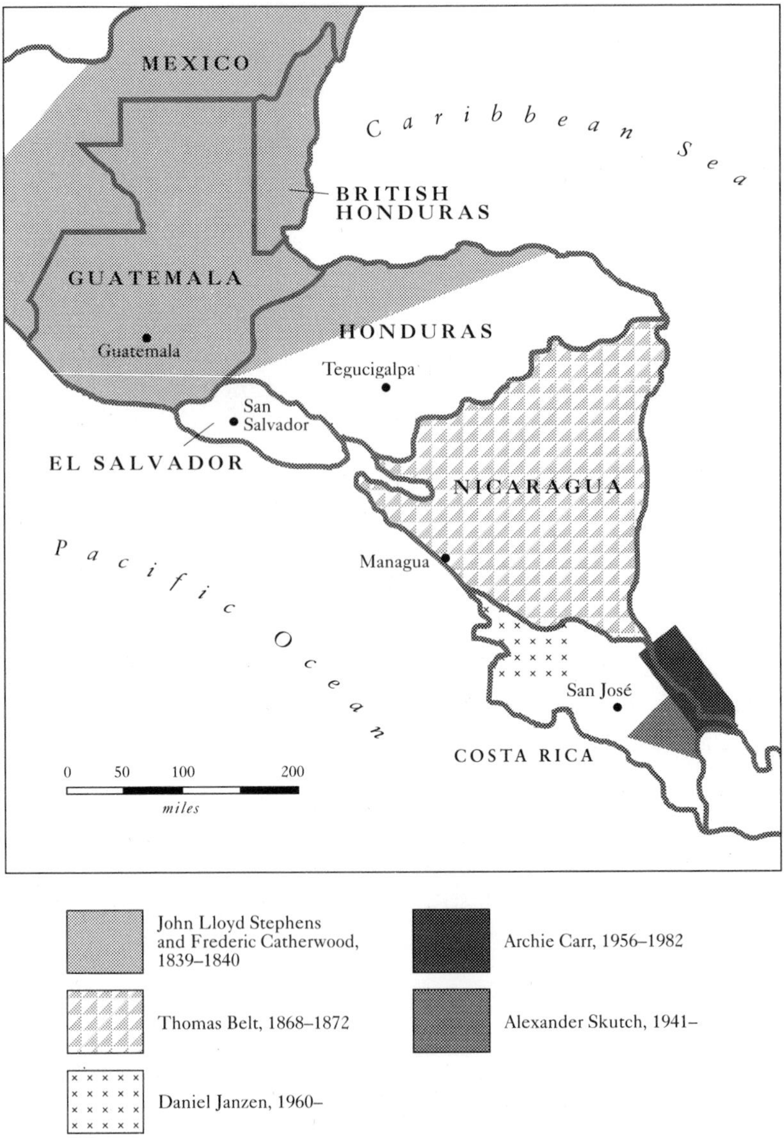

MEXICO
Caribbean Sea
BRITISH HONDURAS
GUATEMALA
Guatemala
HONDURAS
Tegucigalpa
San Salvador
EL SALVADOR
NICARAGUA
Managua
Pacific Ocean
San José
COSTA RICA
0 50 100 200
miles
John Lloyd Stephens and Frederic Catherwood, 1839–1840
Thomas Belt, 1868–1872
Daniel Janzen, 1960–
Archie Carr, 1956–1982
Alexander Skutch, 1941–

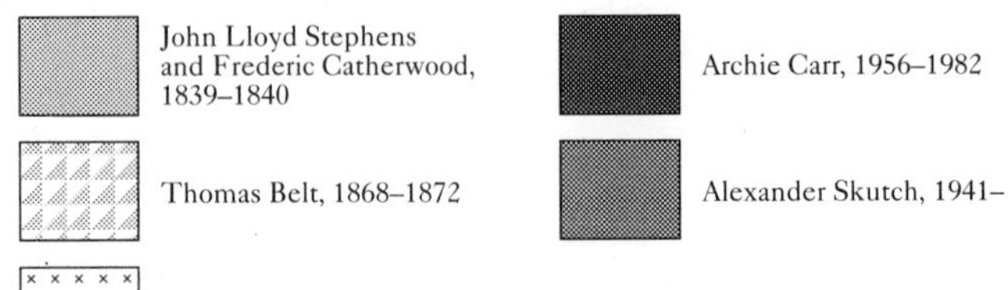

THE EXPLORERS

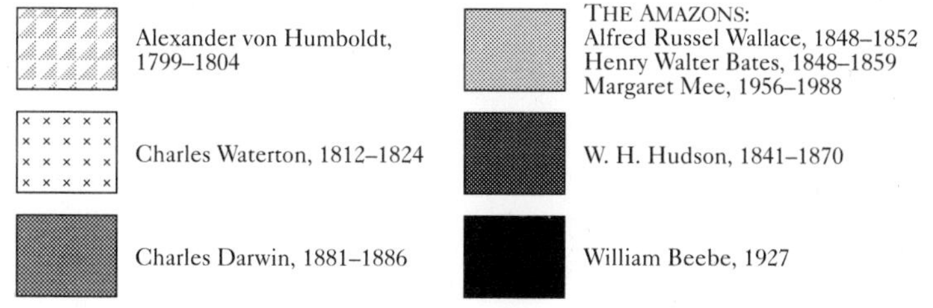

Footsteps in the Jungle

Introduction

THIS BOOK PORTRAYS thirteen of the luckiest men and women of all time: the explorers who first mapped the tropical regions of the Americas and discovered the glorious biological orgy that is nature in the tropics. They were the first to color the chorus of tropical birds, the first to know the swiftness of the jaguar. The first to learn the loves of the orchid family and to collect the daunting variety of moths and butterflies and beetles. They were first to record the swift and haunting settings of the equatorial sun. First to run the rivers Amazon. First to climb the Andes. The first to dive Caribbean coral reefs. And the first to unearth the ruins of America's pre-Columbian civilizations. They were and are the Indiana Joneses of science, plunging ahead often recklessly on the path to discovery — and emerging with fabulous tales of scientific adventure.

Such lives as have been led by Archie Carr, who roamed the tropical beaches of the world studying sea turtles, or Margaret Mee, the botanical painter who made fifteen expeditions to Amazonia, hold a special place in the history of science. They are an antidote to the dismal late-twentieth-century image of bloodless scientists separated from nature, injecting white rats with tumors in a corporate research lab. These jungle explorers remind us that the laws of nature can be found in nature. And they restore in us the sense that beauty need not be neglected in the pursuit of scientific truth.

Of all these tropical trailblazers, only Charles Darwin's name is well known today. Yet along with Darwin these forgotten explorers had a whopping impact, for the American tropics became the Bell Labs of natural law and evolutionary theory. Diversity, variation, adaptation, selection—these were only a few of the ideas that first jumped out of the jungle. And from this American tropical workshop, science emerged democratic in practice and universalist in spirit. Alfred Russel Wallace was an orphan, a dropout, an unemployed surveyor's apprentice whose ambition became to find the key to evolution in the Brazilian rain forests. W. H. Hudson grew up on a ranch on the Argentine pampas and never went to a true school; yet he became the outstanding South American ornithologist of the nineteenth century and the most famous natural history writer of his time. Alexander von Humboldt was a Prussian who went to Spanish America with French measuring equipment and showed that the same physical laws affect all countries equally. That this laboratory for underdogs is still successfully running today is shown by the scientific explorations of Dan Janzen, who is reforesting Guanacaste Province in Costa Rica according to ecological principles he learned in thirty years of fieldwork there.

The twelve men and one woman portrayed here were biologists, zoologists, journalists, mining engineers, architects, textile manufacturers, schoolteachers. Although they come from different countries and different centuries, and practiced different professions, once they hit the ground they all became curious naturalists and indefatigable travelers. All were tough-minded. They had to be. Along their trails they endured physical suffering and personal loss, shipwrecks, malaria, starvation, injuries, infections, and the torment of biting insects. Perhaps worse was enduring the isolation and loneliness that often comes with the territory of exploration. Yet they persevered, if only to risk their journeys of discovery. For all of them, the tuition they paid was worth it, because tropical nature was their great university of life, with an almost infinite library of open shelves to browse. A place to observe the animals and plants, rivers and forests of a New World without limit. And a magical source of images and symbols, insights into life and all its processes.

The jungles they entered weave in and out of historical fact and human fantasy. Ever since Cristóbal Colón left the port of Cadiz in 1492, heading he thought and prayed for the Indies, the American tropics have represented the European's otherness—those parts of himself that he did not, or could not, psychologically accept within, and so projected outside himself onto the big green screen of the New World and its peoples. If the European considered himself civilized—and he did—then the tropics became an area of savagery, barbarism, even cannibalism. If the European bemoaned his sense of original sin—and he did—then the tropics took on for him the aspects of paradise, where naked, innocent, nonmonogamous noble savages lived in Elysian Fields. If old age and death were the burdens the European could not bear, the American tropics would provide the fountain of youth.

For the Spanish hidalgo, armored and mounted on his Andalusian charger, the New World had to provide suitable enemies. For the underclasses, the New World's first word was *opportunity*. The newly discovered lands filled the heads of millions of men held down by feudal bonds with the notion of getting something for themselves and changing their station in life. The European lust for gold soon promoted the idea that in the jungles of America dwelled El Dorado, the Golden Man, who lived in Cibola among the seven cities of gold.

Almost all the new products that came out of the

Americas to whet the appetites of Europe were essentially tropical products: sugar, tobacco, coffee, chocolate, cotton, oranges, bananas, pineapples. In this way, too, the tropics filled a gap in the life of Europe. To the imaginative capitalist adventurer of the seventeenth and eighteenth centuries, the New World tropics symbolized something powerfully compelling—the idea of profits, huge and legendary profits. Profits worth big risks. The early romantic blush of capitalism among the Dutch, the French, and the English was intimately tied to the sudden appearance on the world stage of this vast new area, conceived as essentially vacant and open for business. The New World tropics in the Age of Discovery and Conquest came to serve as a kind of unmarked Swiss bank account of the Old World's collective psyche, where Europeans deposited fears and fantasies, no questions asked.

So intent were the first Europeans on their disparate missions of New World exploitation that the first two and a half centuries after Columbus went by without anyone making the effort to describe objectively what these lands were like, or to look at their flora, fauna, and geography in an orderly way. Expeditions did not carry along natural philosophers, for neither the backers nor the leaders were interested in gaining knowledge of the New World apart from commercial and trade-related information. "All these islands are very beautiful, and distinguished by various qualities," commented a clueless Columbus. "They are accessible, and full of a great variety of trees stretching up to the stars."

Nor were the Europeans interested to know the Neotropics' aboriginal inhabitants before most of them were annihilated by smallpox, measles, influenza, alcohol, or direct liquidation. The tragically slapdash way the first Europeans confronted the only New World they would ever have the chance to discover kept them willfully ignorant of the natives' great storehouse of biological, agricultural, astronomical, medical, and spiritual knowledge. Had the Europeans been interested enough to note these peoples' ways, might we long ago have found cures for cancer, or avoided world wars, or patched the hole in our souls that comes with alienation from nature?

When Alexander von Humboldt traveled to the sources of the Orinoco River in 1800, his primary mission was to disprove once and for all the existence of El Dorado and the Seven Cities of Gold, a legend that had been around since the time of Sir Walter Raleigh. Humboldt traveled fifty-five days upriver in a canoe and discovered the Orinoco's head waters in cataracts it shares with the rivers of the Amazon system. With a bold stroke, Humboldt cut the cord tying the American tropics to the medieval mind-set, and set humanity on the course of enlightened investigation. A curious fact is that the great tropical explorers influenced, above all, one another. Darwin devoured Humboldt before he shipped out for Brazil, Patagonia, Galápagos. It was reading Humboldt and Darwin's *Voyage of the Beagle* that convinced Wallace to escape the misery of industrial England and strike out for the Amazon Valley. Wallace turned up the flame under Henry Bates's desire to traverse the Brazilian jungles and establish new scientific facts. Only Humboldt had no one to influence him: he was the true discoverer of America.

By conquering the tropics with intelligence and bravery, Humboldt and his followers forged a new way of meeting the unknown. Through their memoirs and scientific accounts, the tropical forests left dark and strange by the Age of Discovery and Conquest became vibrant places where humans could nestle up to the warmth of millions of living creatures and the lushest landscapes on earth. A transformation has occurred. Jungles now are places to which people actually feel compelled to travel, to witness for themselves the voluptuous abundance of tropical nature firsthand.

They have become part of humanity's patrimony, an important symbol of planetary unity in the long twilight struggle to preserve a healthy global environment.

The method of these essays is simply to tag along into the field, following the explorers on their voyages and expeditions, offering a companion guide to their discoveries and a context to their imaginative quests. My hope is that such field science as the tropical explorers practiced on the hoof rouses the reader's imaginative juices and kindles dreaming—especially that romantic enthusiasm for tropical adventures that readers can embark upon at first by proxy, perhaps later in reality. Young people especially have always found tropical nature a source of fascination and wonder. It was as a boy sitting in movie theatres on rainy summer afternoons, absorbed in the tales of Bomba or Tarzan, that I was first seduced toward the two dozen or so trips I would later make into the American tropics. It was to lead curious young people across the palm line and through the bamboo curtain, to give them a taste of some of the tropical delights of the Americas, that I originally undertook this book. I confess that several of these chapters began as lectures on various campuses. I am indebted to all the colleges and schools that have allowed me to share my enthusiasm for tropical nature with the coming generation of explorers.

1

Alexander von Humboldt and the Biological Discovery of America

IT MAY HAVE BEEN sheer coincidence that Alexander von Humboldt began his adventures in the New World in 1799, at almost precisely the same spot where Christopher Columbus abandoned his in 1502—near the mouth of the vast, majestic, and still highly underachieving Orinoco River on the coast of present-day Venezuela. Yet this geographic serendipity between the New World's two greatest explorers—one the celebrated discoverer and the other the scientific explorer all but forgotten today—will help us reflect on the meaning of discovery. Was the discovery of America a simple matter of who got there first? Or was it a matter of bold and indefatigable travel, boundless curiosity, meticulous measurement, patient observation, and a passionately romantic perception of the natural and human forces at work in this luxuriant and unruly continent? Let us see.

When Columbus reached the mouth of the Orinoco on his third expedition—the so-called Southern Voyage—he was already an aging mariner with failing health and an arrest record. His search for Cathay was by now making the Admiral of the Ocean Sea obsessive, peevish, even somewhat delusional. The Orinoco is one of several American river systems whose volume of fresh water pouring from the river's mouth is so enormous as to dilute the ocean's salinity for many miles out to sea. The Old World had no experience of these mighty American rivers. Columbus tried to understand the Orinoco according to Christian dogma. He wrote in his log: "The Scriptures tell us that in Earthly Paradise grows the tree of life, and that from it flows the source that gives rise to the four great rivers: Ganges, Tigris, Euphrates, and Nile. The Earthly Paradise, which no one can reach except by the will of God, lies at the end of the Orient. And that is where we are."

In fact, where Columbus was was at sea between two ages. The astrolabe and compass that had helped him reach the American continent were among the tools, developed or refined at Prince Henry's School for Navigation, that launched the Enlightenment. But Columbus's web of beliefs were spun of medieval doctrines. It is important to remember that while Italy was beginning the Renaissance in the quatrocento, Christian Spain was still fighting off the Moslems, who had ruled the Iberian Peninsula since the eighth century. Seven hundred years of religious warfare—the subjugated race in his own homeland—had produced in the Spaniard a warrior mentality, which valued above all else

Alexander von Humboldt plunged into the equatorial jungles with the most complete array of scientific instruments to that time.

might, authority, discipline, hierarchy, and piety. Columbus was the hybrid product of the Italian Renaissance and the fanatical warrior culture of medieval Christian Spain. His was a mind that could not easily accommodate new facts or values, demonstrated by his actions on nearing the landmass of the Americas.

Without attempting to penetrate the Orinoco River, nor even to explore the coastal shoreline, Columbus immediately turned his sails back toward Europe. His immediate intent was to raise a much larger expeditionary force, to return and assault the "Earthly Paradise"—which he understood to mean "the seat of all the world's gold." For gold, Columbus wrote to the Spanish sovereigns, "is the most precious of all commodities; gold constitutes treasure, and he who possesses it has all he needs in this world, as also the means of rescuing souls from purgatory, and restoring them to the enjoyment of paradise."

The power to redeem souls! Gold enough to raise a great army and recapture Jerusalem from the Mohammedan infidels! These were the prophetic signs Columbus read on the tropical features of the green continent he found. In the event, it was the overture to the Conquest.

Alexander von Humboldt landed in the New World at the age of twenty-nine, when life appeared to him a "boundless horizon." The only abnormalities he suffered from were an overabundance of physical energy—something like what we would now call hyperactivity—and an insatiable scientific curiosity that drove him to take tropical America's measurements. Before Lewis and Clark had yet explored the interior of North America, Humboldt faced a southern continent not significantly better known than in Columbus's day. It was not only that large parts of the landmass had never been described by a scientific traveler; the physical structures of the globe itself were either unknown or poorly understood. All existing maps of the American hemisphere, for example, were wrong, based on faulty astronomical readings, making the ships Humboldt took during his five-year excursion invariably off course and several hours late. Nothing would become more typically Humboldtian than to see this high-spirited righter of factual wrongs leaping off a ship in some exotic port, informing the harbormaster that the maps had his harbor's geographic coordinates wrong, and there and then setting up his trigonometric instruments to correct them—before he had even found lodgings.

The science of geology did not exist. Paleontology was primitive. The climate was poorly understood—especially its influence on plant and animal life. In Europe they were still trying to classify American species. The Indian nations of Spanish America, from the Aztecs in Mexico to the Incas in Peru, were not considered a legitimate subject of scientific inquiry, and so the study of antiquaries and ancient sites, known as archaeology, which has yielded so much information on the human past, also did not exist.

The pattern of colonial settlement fell almost entirely on the coastal plains of tropical America, so little was actually known about the huge interior of the Americas in the Tropical Zone. No Europeans had found the sources of the Orinoco, let alone the Amazon. None had climbed the Andes or the many volcanoes of the Pacific Rim. The continent under the Southern Cross lay unexplored. The tropical jungles were only the stuff of myths and legends. European scientific ideas about the New World were based on such misnomers—like the ludicrous notion of Buffon that the heat and humidity rendered American plants and animals inferior to European ones. Buffon himself had never, of course, set foot in the Americas.

The scientific canvas available for Humboldt in tropical America was broad indeed, and he came with a full palette and a master's skills. Born in 1769, the son of a Prussian officer and a cold mother he never got on with,

Humboldt grew up at the height of the Age of Reason and had been educated in, or studied himself, nearly every branch of the natural sciences—from mining engineering (his college major) to physics, surveying, astronomy, botany, mathematics, and medicine. He was pickled in the universalist and encyclopedic ideas of the French philosophers; the democratic principles of Thomas Jefferson; the free-market liberalism of Adam Smith; and the romanticism of his German friends Goethe and Schiller. His interests were eclectic (he had performed, for example, experiments to study electricity by attaching wires to his own body and sending a current through his back muscles), his mind wide open, his love of the natural world absolutely devoted. From earliest youth Humboldt had been such an ardent collector of flowers, butterflies, insects, and so forth that his family sneered and nicknamed him "the little apothecary." As a boy he had studied maps of distant and exotic continents, taking, as he put it, "a pronounced sensual pleasure" in their shapes and forms, filling his being with a classic eighteenth-century wanderlust "to travel to distant regions where Europeans have seldom visited."

Humboldt also brought to America a kit of scientific instruments that could have turned Columbus into Galileo. It was without question to that time the most complete mobile geoscience lab of scientific instruments assembled for measuring, gauging, testing, recording, surveying, mapping, collecting, counting, probing, tickling, sensing, and viewing American nature in all her tropical splendor. The tools to measure America were provided to Humboldt by the Bureau of Longitudes in Paris, France's young scientific agency born of the Enlightenment. According to Humboldt's own published list, they included:

a timekeeper by Lewis Berthoud;

a three-foot achromatic telescope by Dollond ("for the observation of Jupiter's satellites");

a telescope by Caroche ("with an apparatus to fix the instrument to the trunk of a tree, in forests");

a sextant by Ramsden;

a snuffbox sextant by Troughton ("This small instrument is very useful for travellers when forced in a boat to lay down the sinuosities of a river, or take angles on horseback without dismounting");

a theodolite by Hurter;

a quadrant by Bird;

a graphometer by Ramsden ("with a magnetic needle and a wire meridian to take magnetical azimuths");

a variation compass by Le Noir;

a magnetometer of Saussure;

two barometers by Ramsden;

several thermometers by Paul, Ramsden, Megnie, and Fortin;

two hygrometers of Saussure and Deluc ("of hair and whalebone");

two electrometers of Bennet and Saussure ("of gold leaf and elder pith, furnished with conductors four feet long, to collect, according to the method prescribed by Mr. Volta, the electricity of the atmosphere, by means of an ignited substance, which yields smoke");

a cyanometer by Paul ("to give me the means of comparing with some certainty the blue colour of the sky");

a compound microscope of Hofmann;

a land surveyor's chain;

a rain gauge;

a standard meter by Le Noir;

a galvanic apparatus;

reagents ("to try some experiments on the chemical composition of mineral waters");

and "a great number of small tools necessary for travellers to repair such instruments as might be deranged from the frequent falls of the beasts of burden."

Humboldt came at the beginning of scientific accuracy, data collection, repeatable results. He thought the ultimate aim of his observations would be to recognize the underlying principles in the diverse phenomena the natural world offers us—in short, the laws of nature. His instruments packed, on the eve of his departure from Europe Humboldt wrote his friend Freiesleben, "Man must strive for the good and the great! . . . I shall collect plants and fossils, and with the best of instruments make astronomic observations. Yet this is not the main purpose of my journey. I shall endeavour to find out how nature's forces act upon one another, and in what manner the geographic environment exerts its influence on animals and plants. . . . I must find out about the harmony in nature."

Humboldt and his traveling companion, the French botanist Aimé Bonpland, intended first to travel up the Orinoco River as far as they could, to the source if possible. Humboldt explicitly wished to strike a coup against the "ancient custom of dogmatizing geographers," disproving with empirical evidence the story, as old as the Conquest, that the Orinoco began in a great lake, in the midst of which stood El Dorado, the City of Gold.

Before they could move away from the coast of Venezuela, however, Humboldt's curiosity and restlessness got the best of him. He and Bonpland spent three months in a giddy flush of tropical naturalizing. "We run around here like mad," Humboldt wrote to his brother. "In the first three days we couldn't proceed with any scientific work. We would pick up an object and within seconds reject it for a more striking one. Bonpland assured me that he would go stark mad if the excitement didn't stop soon."

Everywhere he turned in those first weeks of arrival, Humboldt saw tropical America with love at first sight. So profoundly did the tropical landscape move his soul that Humboldt soon felt as if the trees and flowers had washed his mind clean. He wrote, "There is something so great, so powerful, in the impression made by nature in the climate of the Indies, that after an abode of a few months we seemed to have lived there during a long succession of years. . . . Between the tropics . . . everything in nature appears new and marvelous. In the open plains, and amid the gloom of forests, almost the remembrances of Europe are effaced; for it is the vegetation that determines the character of the landscape, and acts upon our imagination by its mass, the contrast of its forms, and the glow of its colours."

Before setting out on the Orinoco, they had obtained specimens of sixteen hundred plants, six hundred of them species new to science. They had witnessed a fascinating meteor shower, recorded an eclipse. Humboldt was busy all day asking questions of everyone he encountered. He could soon speak Castilian better than the locals. At night he was out under the stars, taking coordinates. On the 4th of November the devil himself obliged Humboldt by offering up an earthquake to record:

"About two in the afternoon, large clouds of an extraordinary blackness enveloped the high mountains of the Brigantine and Tataraqual. They extended by degrees as far as the zenith. About four in the afternoon, thunder was heard over our heads, but at an immense height, without rolling, and with a hoarse and often interrupted sound. At the moment of the strongest electric explosion . . . there were two shocks of an earthquake, which followed at fifteen seconds distance from each other. The people in the streets filled the air with their cries. Mr. Bonpland, who was leaning over a table examining plants, was almost thrown on the floor. I felt the shock very strongly, though I was lying in a hammock."

There was so much to see and learn, it's a wonder they got away at all. But in February 1800 the Age of Reason's master explorer and his companion Bonpland left the coast and set out on an epic journey, first across the heavily forested coastal mountain ranges and the elevated grassy plateaus of the *llanos*; then up the Apure River to the confluence with the Orinoco; upstream again on the river's southern tributary the Atabapo, and all the way to the watershed shared by the Rio Negro, one of the Amazon's chief tributaries. They then navigated up the Rio Negro to within two degrees of the equator to the confluence of the Rio Casiquiare.

Humboldt plunged into these equatorial jungles in one of the least-known and most hazardous regions of the Americas; braved torrential rains and the torture of mosquitoes; made friends with the natives, learned their languages, and recorded the first documented instance of geophagy (the eating of dirt) among them. Humboldt and Bonpland lived for weeks on nothing but manioc roots, bananas, and occasional monkey meat. They had to portage their boat and gear for days and days through jungle thickets. But one evening in May their canoe slid past the steep river banks and masses of driftwood where the Rio Casiquiare joined the Orinoco again, and Humboldt ascertained that the great river Orinoco, which had flipped Columbus, actually shared waters with the mother of all American rivers, the Amazon. "In an uninterrupted navigation of 920 miles," Humboldt reported in *Aspects of Nature*, "I passed through the singular network of rivers, from the Rio Negro, by the Casiquiare, in the Orinoco; traversing in this manner the interior of the Continent, from the Brazilian boundary to the coast of Caracas."

When Humboldt returned to Caracas seven months later, he had covered altogether 6,443 miles, mostly in native dugout boats known from one end of South America to the other as *piraguas* (and in the Louisiana bayous as pirogues). In addition to mapping the Orinoco and verifying her sources, he had taken magnetic readings at latitudes close to the equator, where he found that the earth's magnetism differed greatly from that in northern lands. Humboldt had been asked by the Bureau of Longitudes to take along newly developed instruments for determining changes in magnetic field. These assays would lead to Humboldt's proposing the law of declining magnetic intensity between the earth's magnetic poles, one of many pieces of fundamental geophysical knowledge he discovered.

But by any measure, the richest scientific bounty of Humboldt's Orinoco expedition was the collection of twelve thousand plants he and Bonpland brought back. The sheer abundance and variety of life forms in the tropical forests of America made a profound and lasting impression on Alexander von Humboldt. The combination of high humidity and high temperature produced, far from the decadent life forms suggested by Buffon, a magnificently diverse community of plants and animals. Humboldt compared the flora with the northern forests:

"In the temperate zone . . . forests may be named from particular genera or species, which, growing together as social plants, form separate and distinct woods. In the northern forests of Oaks, Pines, and Birches, and in the eastern forests of Limes or Linden trees, usually only one species . . . prevails or is predominant. . . . Tropical forests, on the other hand, decked with thousands of flowers, are strangers to such uniformity of association; the exceeding variety of their flora renders it vain to ask of what trees the primeval forest consists. A countless number of families are here crowded together and even in small spaces individuals of the same species are rarely associated."

Among these green multitudes, Humboldt saw with his own eyes the amazing exuberance and vitality of tropical forests. Shortly after his return from the Orinoco expedition, he wrote to his friend Willdenow in Berlin: "We were barely able to collect a tenth of the specimens met with. I am now perfectly convinced of [the] fact . . . that we do not know 3/5 of all the existing plants on earth!"

The diversity of tropical forests is so taken for granted nowadays that we can hardly imagine it as something requiring discovery. Yet Humboldt was the first to observe and record it in a rigorous way. Equinoctial America had what the Enlightenment's natural philosophers called the "life force" (*lebenskraft*)—only it was not some kind of real current, like electricity, as Humboldt expected, but was rather expressed in the wild, reckless proliferation of tropical species. Millions of plants crowding toward the sun, launching carpets of flowers, dropping bushels of fruits. The many kinds of brilliantly plumed birds. The magnificent, almost uncanny, variety of butterflies and insects.

This evidence of tropical *lebenskraft* made Humboldt see that the basis of the elaboration of differences was actually in the uniform circumstances of the Neotropics' physical geography. His understanding of the environmental unity underlying biological diversity was the very first glimmer of the science of ecology, an integrative science that seeks to understand the parts of nature in relation to the whole. How very different was the meaning of discovery to Alexander von Humboldt and Christopher Columbus!

Humboldt found other important changes of meaning in the New World. In Europe the word agriculture meant the growing of crops for human consumption. In tropical America agriculture had come to mean the growing of single crops for export to earn cash—sugar in the West Indies, cotton in the American South, coffee and tobacco at higher elevations, and so on. From the days of Columbus's search for treasure, America had evolved as the economic continent, conquered and exploited for the explicit purpose of producing wealth.

Humboldt found that the way wealth was produced was itself transformed in the New World. In the North of the United States, an experiment was taking place in free labor, agrarianism, and political democracy. But below the celestial equator, a new system had developed for the forced exploitation of human labor: enslaved natives in the gold and silver mines, imported African slaves on the plantations.

Having missed Cuba on the inbound trip because of a typhus epidemic there, Humboldt and Bonpland went to Havana to spend the winter of 1801. They traveled all through the island's interior, studying the sugar plantations and the conditions of slave labor. By and large Humboldt avoided direct criticism of social and political institutions. After all, he was traveling under the auspices of the Spanish crown. Before arriving in the isle they called "Pearl of the Antilles" he had heard of the benevolence of the Cuban slave codes. The crown and the church were to guard the rights of the slaves, while slaves were permitted to join the Catholic church and take the sacraments. By the turn of the nineteenth century, slaves under Spanish rule generally had the right to marry, own property, appeal to magistrates if cruelly punished, and were accorded legal access to manumission.

Humboldt acknowledged that slavery in the city of Havana was ameliorated by these codes. But he was not one to mistake the life of the urban bondsman for that of the plantation field hand, and he immediately set forth from Havana to visit every sugar plantation he could find on the island. On the plantations Humboldt gathered information from slaveholders and slaves alike. He found the moderate Spanish slave codes went unenforced—and therefore disregarded. The Cuban sugar planters abused their slaves as severely as did the Americans and the British. During the sugar harvest they were worked incessantly day and night. Masters ignored the prohibition against Sunday and holiday labor. The right to marry was nullified, Humboldt found, because the planters were disinterested in breeding slaves. Ratios of fifteen or twenty male slaves to one female were not uncommon. One plantation he visited had seven hundred males and no females.

Slaves were punished with the lash, stocks, iron collars, leg chains, pins, and knives. Runaways were recaptured with dogs and whipped to the point of death as an example to others. The planters had made an economic decision: as long as the slave trade from Africa continued, it was cheaper to replace field hands after a few years with fresh labor than to maintain their strength by humane treatment. In short, it was more profitable to work slaves to death.

The sight of such abuses of humanity roused Humboldt to "indignation and fury." But as a student of human nature he thought that those acting immorally out of economic motives would never be moved by moral suasion alone. Instead of preaching on humanistic grounds, Humboldt responded with a brilliant and crisp economic counterpunch in his *Political Essay on the Island of Cuba.*

In Cuba, Humboldt employed for the first time the "totalizing" method of political geography, first suggested by Montesquieu and elaborated by Thomas Jefferson in his *Notes on the State of Virginia.* It meant using statistics to analyze human society—the nascent science of economics. Using data on slave and white populations, crop production, commerce, revenues, taxes, and trade balances with great dexterity, Humboldt endeavored to prove in his essay that the slave system in Cuba and the Caribbean was leading inexorably to economic instability and political revolt.

Humboldt argued that the Caribbean Basin possessed the climate, the natural resources, and the trade routes to become a great "cradle of civilization"—equal to the Mediterranean. The question was how to get there. Monoculture—with its slave plantation economy? Or a free market made up of educated free citizens? Humboldt the social scientist showed that the plantation system was in fact undercutting the whole region's future development, at the same time threatening to ruin the very planter class that was the beneficiary of human bondage. It was in the planters' own best economic interest, Humboldt argued, to end slavery.

But his analysis was not heeded. Cuba, Haiti, and the rest of the Caribbean and Latin America plunged into precisely the kind of political instability and lack of development Humboldt had forecast. He remained adamant on the slavery issue his entire life, a forceful advocate of ending the slave trade with Africa and a passionate defender of the equality of Negroes. He later wrote, "It is for the traveller who has been an eyewitness of the suffering and degradation of human nature to make the complaints for the benefit of the oppressed."

Having shipped their botanical collections back to Europe on the chance they might not survive their tropical adventures, Humboldt and Bonpland prepared to plunge deeper into the unknown continent. In April 1801 they shipped out for the coast of New Granada, present-day Colombia. Disembarking at the mouth of the Magdalena River, they canoed fifty-five days upstream to the foothills of the eastern Andes, then trekked overland by mule train to Bogotá.

Along the way Bonpland grew seriously ill with headache and nausea. But by now the fame of the Prussian explorer had spread so wide that a cavalcade of Bogotá's leading citizens dressed as Spanish knights came out to greet the two men and escort them into town. A visit from Humboldt had become a kind of stamp of legitimacy to Spanish colonials with social inferiority complexes. Everyone wanted to meet the great explorer and man of science. Everyone wanted to see the beautiful new maps he was making; to admire his exquisite landscape drawings; to examine his telescopes, his newfangled barometers, the trunks and trunks filled with natural curiosities. More than anything, the leading citizens of Bogotá came simply to see the famous explorer Alexander von Humboldt: he had become South America's first celebrity.

With regal hospitality in Bogotá, Bonpland recovered his health while Humboldt went off with locals to hunt for mastodon bones. He found them. From Bogotá the two men struck off through dense bamboo thickets and uninhabited forests of lofty wax palms, their white trunks rising from dense floors of tree ferns. They were now climbing into the central Andes for what would be nearly two years of crisscrossing that *cordillera*, the spine of mountains and rim of volcanoes along the Pacific side of South America.

This journey would yield an even richer scientific harvest than the Orinoco expedition. At every point Humboldt tirelessly measured, collected, surveyed, took readings, notated, mapped, sounded, and drew landscapes. At every turn he sought positive knowledge with that special Cartesian craving to understand how things work that we now associate with the late eighteenth century. Many years later, one of the hundreds of local Indian guides Humboldt employed during his travels told a biographer that this German, who spoke flawless Spanish and numerous Indian dialects, was not as intelligent as everyone seemed to think. Otherwise why did he have to keep asking so many questions about things that every simpleton knew—the names of rivers and mountains and plants? Probably due to a weak memory, the informant said, Humboldt had to write everything down in a small book.

In the Andes, Humboldt was able to realize his long-held ambition to study how the geographic environment influences the life of plants. He observed that plant forms require a particular combination of temperature, rainfall, and soil to flourish. These parameters fluctuated, Humboldt found, according to their altitude and geographic location. He gathered sufficient data to show that the geographic distribution of vegetation was conditioned by the annual temperature range at different elevations above sea level. The geography of plants, Humboldt said, was "intimately associated with the study of the distribution of heat over the surface of the earth."

Different plants and plant types increased or decreased "as we recede from the equator towards the poles." Humboldt mapped these vegetation zones in which the latitudinal range of flora repeated itself vertically on mountain slopes rising from sea level to snowy summits. He proposed a system of demarcations called isothermal lines: "The curves of the isothermal lines . . . correspond with the limits which are seldom passed by certain species of plants, and of animals which do not wander far from their fixed habitation, either with respect to elevation or latitude." This was the first geographic theory of vegetation, highly influential in helping Charles Darwin think about the adaptation of species to physical factors in the environment. Isothermal maps showing vegetative zones also became one of Humboldt's most marketable scientific contributions, useful in making growing-zone charts for farmers, foresters, agronomists, and fruit growers.

Arriving in Quito, Ecuador, in January 1802, Humboldt began to explore the region's numerous live volcanoes. He continued his study of vulcanism a few years later when he and Bonpland spent a year traveling in Mexico. Humboldt climbed their peaks, analyzed their gases, charted their geologic composition, studied their physical structure, and, of course, noted their precise height and geographic coordinates. Always superb at extrapolating general principles by connecting facts collected in different and widely separated locations, Humboldt noticed that while the whole eastern part of the Americas was without fire-emitting mountains, all the American volcanoes were on the side of the continent opposite Asia, aligned on a belt extending some 7,200 miles from Mexico to Chile. In *Aspects of Nature* he wrote, "These assemblages of volcanoes, whether in rounded groups or in double lines, show in the most conclusive manner that the volcanic agencies do not depend on small or restricted causes, in their proximity to the surface of the earth, but that they are great phenomena of deep-seated origin."

The arrangement of volcanoes along this rim of fire, Humboldt claimed, could only mean that the liquid magma rose on a single line of weakness in the earth's crust. This discovery was yet another fundamental clue in studying the earth's geologic formation and the fault lines—later in working out the theory of plate tectonics.

What could Humboldt not accomplish when he devoted himself to a particular problem or project? In June 1802 he and Bonpland set their sights on ascending Ecuador's extinct volcano Chimborazo, then thought to be the highest mountain in the world. Without the aid of down parkas, freeze-dried food, or bottled oxygen, the climbing team made up of Humboldt, Bonpland, a local naturalist named Carlos Montufar, and one Indian headed for the summit:

"With extreme exertion and considerable patience, we reached a greater height than we had dared to expect, for we were constantly climbing through the clouds. In many places the ridge was not wider than eight or ten inches! To our left was a precipice of snow whose frozen crust glistened like glass. The angle of this icy slope was thirty degrees. On the right lay a fearful abyss, from 800 to 1000 feet deep, huge masses of rocks projecting from it. . . . At certain places where it was very steep, we had to use both hands and feet, and the edges of the rock were so sharp that we were painfully cut, especially on our hands . . . one after another we began to feel sick from nausea and giddiness."

Finally they were stopped in their tracks by a chasm some 400 feet deep and 60 feet wide, filled with soft snow. "We set up the barometer with great care," Humboldt reported. "Air temperature was three below freezing. . . . According to the barometric formula given by Laplace, we had now reached an elevation of 19,286 feet." Humboldt had become the world's highest human. It would be another thirty years before British surveyors reached the Himalayas.

In Mexico Humboldt drew the first accurate map; wrote an exhaustive technical report on reform of the silver mining industry; made the first geologic sections—the fundamental analytic tool in geology; and delivered a series of brilliant lectures on stratigraphy, which helped scientize the search for minerals and oil. And he began relentlessly to collect information on the Indian ruins and to gather economic and demographic statistics for an analysis of Mexico's "caste" system under Spain—a second foray into economic geography, later published as the *Political Essay on the Kingdom of New Spain.*

In Peru he observed the transit of Mercury; helped gauge the earth's speed of revolution round the sun; and gathered astronomical information that later helped calculate the ellipsoid shape of the earth's orbit. Humboldt fixed the magnetic equator at 7 degrees 27 minutes south latitude. He shipped samples of bird guano back to Europe for analysis, and when they were found to contain nitrogen and phosphorus with thirty-three times the fertilizing capacity of manure, Humboldt gifted the world the fertilizer industry—he never made a single pfennig from it. He was also first to survey the cold current flowing off the Peruvian coast that is named for him, measuring its temperature and rate of flow.

Humboldt was the first to look at ancient Inca roads

with the appreciative eye of an engineer. He found evidence of a vast network, with roadside shelters, servants' quarters, and baths. The roads were constructed over stone foundations out of neatly hewn volcanic rocks, cemented together by gravel, as if built with macadam. To overcome steep slopes, long flights of steps were constructed at intervals (the Incas had no beasts of burden, consequently no wheeled vehicles).

Entering old Inca territory, Humboldt also had the chance to see the few sixteenth-century hieroglyphic native codices that survived the Conquest. The discovery of these manuscripts, he wrote his brother, "revived in me the wish to study the early history of the aborigines of these countries." The zodiac signs the Incas and Aztecs employed on their calendar stones, he observed, in many cases corresponded to Asian symbols. Both the Chinese-Tibetan and native American systems used tiger, snake, dog, hare, and bird. To Humboldt, the calendar stones furnished a clue to the origins of the aborigines. On this basis he proposed the Asian origin of American Indians—all the more astounding since he had no way of knowing that in fact a land bridge had emerged between Siberian Asia and Alaskan America in the Ice Age, making possible human migration into the Americas.

Humboldt thought the study of active languages would eventually prove the Asian connection: "I regard the existence of a former intercourse between the people of western America and those of eastern Asia as more than probable, though it is impossible at the present time to say by what route and with which tribes of Asia this influence was established. . . . We know as yet too little of the languages of America to renounce entirely the hope that amid their many varieties some idiom may be discovered which has been spoken with certain modifications in the interiors of Asia and America."

Humboldt became a stout proponent of the civilizing achievements of the native Americans as well as their natural rights as humans at a time when Europe, and America too, viewed American aborigines as either brute barbarians or noble savages. "A darker shade of skin color is not a badge of inferiority," Humboldt said. "The barbarism of nations is the direct consequence of oppression by internal despotism or foreign conquest."

In short, Alexander von Humboldt was a one-man Academy of Natural Science, happy to share everything he knew, talking at ninety miles an hour on six or seven topics at once, in a running mixture of German, English, French, and Spanish, never pausing once for breath. Humboldt knew so much about the Americas that President Jefferson invited him to Washington to sample his knowledge. Jefferson's Treasury Secretary Albert Gallatin remarked after an afternoon in his company that you learned more in two hours with Alexander von Humboldt than you could learn in a month spent in a library.

A story—which may be apocryphal—is that Humboldt brought a certain live parrot back to Europe because it had once belonged to the last Indian of a certain tribe. It was said that Humboldt managed to write a lexicon of the extinct tribe's language by coaxing the vocabulary out of the parrot word by word!

After five years exploring the tropical regions of America, Humboldt returned to Europe and spent the remainder of his long life publishing the massive results of his investigations in tropical America in thirty volumes. It was to be a new form of literature, a type of atlas in prose with separate volumes of maps, dictionaries, scientific engravings, and so forth, covering every aspect of the natural world—obviously influenced by Diderot's *Encyclopedia.* Humboldt never married or had a family, and paid for the entire publishing venture himself, at length spending his entire inheritance. He poured all his accumulated knowledge into his great masterpiece, the full title of which is *Personal Narrative of Travels to the Equinoctial Regions of the New Continent During the Years 1799–1804*. He made good on his goal to describe America's physical structures and how they influence living things. During his lifetime he was the preeminent authority on America and probably the second most famous man in Europe after Napoleon. His reputation as a scientist was nonpareil. He was, in fact, the first to make scientific endeavor an international—and supranational—undertaking. Perhaps the Humboldtian idea that science has no national borders was his most important achievement.

Humboldt's influence on the succeeding generation was monumental—Charles Darwin decided to pursue natural science after reading Humboldt, and carried an edition of Humboldt with him on the *HMS Beagle*. Alfred Wallace decided to explore the Amazon after reading Humboldt. The American John Stephens was inspired to explore Mayan ruins and develop the field of American archaeology by Humboldt's volume on native American antiquities. William Prescott was inspired by Humboldt to write *The Conquest of Mexico*.

So many of Humboldt's scientific insights and discoveries have become part of our general knowledge of the world that they are no longer even associated with him. How many gardeners, looking at the growing-zone chart in a seed catalog, realize that such charts owe their creation to Humboldt's work on isothermals? How many farmers, spreading fertilizer on their fields, know that the genesis of this technique lay in Humboldt's shipment of South American guano? Millions know the name of Humboldt's Current—but don't know who discovered it. It is one sign of this universal man that the force and range of his discoveries have been so integrated into the matrix of modern consciousness.

Today's tropical America, on the contrary, shows too little of Humboldt's social and economic ideas. The rich tropical ecosystems Humboldt first recognized are being wantonly destroyed to profit cattle, timber, and other extractive industries. Tropical forests are being lost at such a rate that millions of species will vanish before we even know what they are, what they do, or what their exact influence in aggregate is on our planet's atmosphere and climate. Nowhere in tropical America has the great new civilization Humboldt envisioned as the "Mediterranean of the Antilles" come into being. Latin American oligarchies could not bring themselves to make economically rational decisions when it came to Indians, ex-slaves, and poor peasants. As a result, Latin American economy is based more than ever on cash crops for export—cocaine being only the most notable. Countries like Mexico must actually import food, while most Latin American populations are worse off nutritionally now than before Columbus arrived. Polarized into separate societies of rich and poor, the continent is mired in its inability to develop and progress. The economic continent Humboldt discovered has unfortunately become the continent of inequality.

In his final work, *Cosmos*, written when he was in his eighties, Humboldt wrote of the link between region and culture, between environment and human history. "The influence of nature's physical traits on the moral nature of people, the secretive mutual interaction of sensual and super-sensual," he wrote, "lends to nature studies a special challenge as yet unappreciated." In his travels through the Americas, Humboldt recognized many of the natural laws that unite our planet. Underlying such knowledge was his view that nature operates essentially by cooperation. Wasn't tropical America one big example of a community made harmonious through the orderly operation of natural forces?

Humboldt's positivist paradigm of nature, however,

lost its appeal to a world emerging from Conquest and colonialism, only to rush headlong into nineteenth-century industrialism and imperial conflict. That world did not wish to hear the harmony of Humboldt any more than the harmony of Mozart. That world took its challenge from a paradigm of nature whose political geography crossed the twin peaks of competition and the mordant struggle for survival. So social Darwinism won its dominant place in the common consciousness of the modern world. Yet under the relentless bludgeonings of history, we can still say with pride that Alexander von Humboldt discovered the true heartbeat of America.

2

Charles Waterton's Wild Wanderings in South America

IN 1820, on his third expedition into the tropical forests of the Guianas, the singular squire Charles Waterton was intent on fulfilling a long-held ambition: to capture a cayman, or New World crocodile. He wanted the reptile for his stuffed zoological collection in the chilly halls of Walton in Yorkshire, England, his family estate. He might have shot a cayman quite easily any day on the Demerara River, where he wandered and collected frequently in the early part of the nineteenth century. But shooting any old cayman, any old way, would never do for Waterton. The "Squire," as he was known to everyone, was nothing if not a perfectionist when it came to one of his favorite avocations, taxidermy. He was not about to settle for a cayman skin damaged by bullet holes, gashed by cutlass, or stabbed by lance, he tells readers in his memoir, *Wanderings in South America*. Nothing but a *live* cayman taken under his personal command would do. And not a puny cayman, either. Only a specimen "of the large kind" would satisfy.

So Waterton readied his canoe in Georgetown, the present capital of Guyana (the former British Guiana), and set off up the Essequibo River with the sun flaming down on his exposed feet. No Charles Waterton expedition could begin, progress, or end without at least one horrendous injury, illness, or accident; its diagnosis, course, and treatment were always described by Waterton in the precious detail most modern writers save for the carnal act. No other naturalist ever recounted so fully the various ill effects of the bites of chiggers and ticks, fleas, red ants, and mosquitoes. No other explorer ever dared impose on his audience page after page describing his fevers, infections, swellings, chills, bruises, abrasions, sprains, and headaches. But then again, few human beings ever had the infinite capacity for getting into weird accidents or acquiring exotic illnesses as did Charles Waterton. Part of this unique talent came from his absolute refusal, on principle, to wear shoes or a hat in the jungle. And part, too, from his habits of sleeping in wet clothes, acting as his own doctor, and, last but not least, showing off.

The cayman hunt started badly. After five nights fishing with a baited shark-hook failed to entice a cayman, Waterton realized something was materially wrong and promptly fired his Negro guide, who, he reported, was "taking on airs" anyway, which Waterton found "intolerable . . . in any expedition where I am commander."

Instead the Squire asked local natives for their advice on capturing a cayman, which they readily agreed to give. As he lay in his hammock beside the river that night, Waterton

Charles Waterton pictured in one of his most famous stunts: riding an alligator on the banks of the Essequibo River.

had one of those epiphanies that often seemed to strike him when prone, swinging, and suspended between two trees. If Montaigne could say in his essays that he always did his best thinking on horseback, Charles Waterton could rejoin that his best thoughts often came to him while swinging in a jungle hammock:

"I considered that as far as the judgment of civilized man went, everything had been procured and done to ensure success. We had hooks and lines, and baits, and patience; we had spent nights in watching, had seen the cayman come and take the bait, and after our expectations had been wound up to the highest pitch, all ended in disappointment. Probably this poor wild man of the woods would succeed by means of a very simple process; and thus prove to his more civilized brother that, notwithstanding books and schools, there is a vast deal of knowledge to be picked up at every step, whichever way we turn ourselves."

This, too, was characteristic of Charles Waterton: on the one hand, his high regard for the knowledge of the locals, a respect that allowed him to travel wherever he liked with Indian aid; on the other hand, his devotion to personal observation in the field as the best way to learn the secrets of tropical nature. Waterton was a practical, hands-on, seat-of-the-pants naturalist, deeply contemptuous of academics and their theories. His was a do-it-yourself approach to natural history. He refused to believe unless he saw with his own eyes. Obviously this approach had its limitations in the dawning scientific age of experiments with repeatable results. But while skeptical of everyone else, Waterton insisted that *he* could be believed. After all, he was an Englishman, a gentleman—a Catholic! But even in his own day most people disbelieved his adventures in the tropics, and his *Wanderings* was put down as the work of a crackpot. It stung Waterton to such an extent that he spent a good deal of his later life engaged in personal vendettas against those who had insulted him in print or in his imagination.

Waterton had been sinking his bait for the cayman, with no success. The next day the Indians tied a packet of four hard wooden hooks, barbed at both ends, to the end of a rope, baited it, and suspended it over the river. About half past five in the morning, the Indians' shouts brought Waterton and his crew on the run—the others ran ahead of the Squire by two minutes, he informs us, because they weren't wearing trousers.

The barbed hooks had done their work. A fierce ten-and-a-half-foot alligator was now held fast to the end of the rope. "Nothing now remained to do," said Waterton, "but to get him out of the water without injuring his scales."

This was easier said, of course, than done. When the Squire told his helpers that he intended to draw the cayman slowly out of the water and then secure him, they looked at one another in disbelief, squatted on their hams, and told Waterton to do it himself.

Waterton's sidekick, Daddy Quashi, a kind of black Sancho Panza whom Waterton sometimes chased around the jungle waving his machete to stir him to action, was for shooting the cayman. Like his Manchegan role model, the artfully named Daddy Quashi was always in favor of an earthy, sensible approach: you give poisonous snakes wide berth; you don't go swimming in piranha-infested waters; you shoot alligators.

The Indians proposed firing a dozen arrows into the reptile to disable it. In short, as Waterton summed up the situation, "They wanted to kill him, and I wanted to take him alive."

Waterton paced the sand bank of the Essequibo, turning over a dozen strategies in his head while Daddy Quashi and the Indians held the alligator in the water by the rope.

At length the Squire struck on a strategem. He would go down on one knee in front of the crew, holding the mast of the canoe sail in front of him like a bayonet fixed to the end of a rifle. When the cayman charged open-mouthed from the river, Waterton would shove his makeshift weapon down the cayman's throat, dispatching the gator without damaging the pelt. Even such a typically screwball plan might have worked had Waterton stuck to it. But he rarely stuck to anything for long. He was that happiest of all hyperactives, the freewheeling jungle explorer, distracted by the distractions of his own distractability.

When the crew understood that Waterton himself would take his stand between them and the cayman, they finally agreed to haul the monster out of the water. The Squire aimed his sail mast. The Indians tugged the line. The alligator came to the surface, then plunged furiously, giving Waterton only enough of a glimpse "not to fall in love at first sight." He ordered the crew to heave away. The alligator arrived on shore in what Waterton called "a state of fear and perturbation."

Did Waterton thrust his wooden weapon? Certainly not. "I instantly dropped the mast, sprang up, and jumped on his back, turning half round as I vaulted, so that I gained my seat with my face in a right position. I immediately seized his forelegs, and by main force, twisted them on his back; thus they served me for a bridle."

It was a quintessential Charles Waterton stunt: hop onto an alligator's back and ride him like a buckaroo. The crew shouted and cheered so loudly the Squire was afraid the cayman would retreat into the river, which would have left him in a fine pickle. So Waterton directed that cayman and rider be dragged forty yards inland. There, after a protracted struggle, they managed to tie the creature up and haul him back to camp in the bottom of a canoe. After breakfast Waterton cut its throat and dissected the cayman.

There was no photography to record Waterton's ridiculous ride, but an engraving was later made for his book. It shows Waterton in the saddle of the reptile with his hair standing on end, as though a thousand volts were going through him. With his inability to keep a straight face, Waterton quipped, "Should it be asked how I managed to keep my seat? I would answer—I hunted some years with Lord Darlington's fox-hounds."

Impetuous, inflammatory, indefatigable, terminally restless, maddeningly obstinate, by turns whimsical and self-important, self-deprecating and self-righteous, temperamental yet unfailingly cheerful, highly irregular, always surprising, and, all in all, way over the top. These are some of the words that come to mind to describe him. There was only one real and original Charles Waterton, though he often seems a made-up character in his own memoirs—peculiarly American in his love of exaggeration, his relish of the whopper, his telling stories on himself—not a little unlike, for instance, Ronald Reagan.

In true pioneer fashion, Waterton went where no white man had gone and invented himself as a kind of Davy Crockett of the tropical frontier. But as an oddball member of the British landed gentry, the stuffiest and by all odds the most sedentary social class in human history (prior to the American suburbanite), Waterton came to define not the tropical pioneer but the nineteenth-century English eccentric: the individualist, true only to his own creed, caring nothing for conventions or acceptance, popular values or fashions, and concerned exclusively with the inner satisfaction of accomplishing his own goals.

Waterton the eccentric, for example, never slept in a bed. He thought it was healthier to sleep on the ground, which he declaimed in an essay titled "On Fresh Air." He awoke each day at 3:30 in the morning. His habit of going barefoot time after time resulted in lacerations, sprains, sun poisoning, and, worse, another opportunity to doctor himself, a lifelong Waterton passion.

Deep in the Essequibo jungle, Waterton fell into hot pursuit of a redheaded woodpecker and soon was so absorbed in the chase that he forgot to look where he was going. He trod on a hardwood stump which, though "only a little over an inch high," punctured "the hollow part of my foot, making a deep and lacerated wound there." The Squire fell to the ground, waiting until the first wave of pain and nausea passed. Then, "I allowed it to bleed freely, and on reaching headquarters washed it well and probed it, to feel if any foreign body was left within it. Being satisfied that there was none, I brought the edges of the wound together and then put a piece of lint on it and over that a very large poultice, which was changed morning, noon, and night."

Luckily, as Waterton put it, there were a couple of cows in the vicinity. And since, as he reasoned, heat and moisture were the two principal virtues of a poultice, "nothing could produce those two qualities better than fresh cowdung boiled." Almost unbelievably, Waterton not only failed to infect the wound, but three weeks after he began his cowflop poultice treatment, "only a tiny scar remained."

Later in life, after his wanderings in South America were over, the Squire used to greet visitors at Walton Hall by running out on all fours like a dog, growling, barking, and biting their legs. Alternatively he would come dancing down the long driveway barefoot and hatless in a snowstorm to greet his visitors. Once, his identification with birds led him to the conviction he could fly. He built a pair of wings and launched himself off the roof of Walton Hall. Another time, striding to the steamer at Dover for a trip to the Continent on a winter's night, he took a wrong turn and walked off the dock, falling headlong into the icy sea. Undertaking his own cures for his many contretemps, Waterton obstinately dosed himself with purgative potions potent enough to kill an ox. His near addiction to self-medication by bloodletting, which he liked to call "tapping my claret," was rivaled only by his genius in the unrecognized art of climbing trees, which he continued to practice daily as an octogenarian.

And there was his almost mystical connection with animals, which, for example, permitted him to handle venomous snakes and constrictors all his life without once being bitten or squeezed. Waterton was convinced that snakes were basically amiable, sluggish creatures that would never purposely attack a man who approached them slowly and with absolute self-command. For one whose life was filled with pratfalls that went awry, Waterton's snake-handling bordered on the miraculous.

Now, everyone who has wandered in tropical America has a snake story or two. But Waterton's beat all. One day while out in the jungle, following a new species of parakeet, the Squire observed a ten foot "coulacanara" snake slowly moving along a timber-cutter's path. This was probably the *Boa imperator*: "There was not a moment to be lost. I laid hold of his tail with the left hand, one knee being on the ground; with the right I took my hat, and held it as you would hold a shield for defence."

The boa was not happy about this turn of events. It pivoted instantly and raised its head three feet off the ground, "as if to ask me, what business I had to take liberties with his tail." Waterton let the snake, hissing and open-mouthed, come within two feet of his head. Then he punched it in the jaws. The poor boa was stunned; Waterton easily seized it by the throat, allowed it to wrap itself around his body, and "marched off with him as my lawful prize." Just then Daddy Quashi arrived on the scene. As soon as he saw the boa he turned tail and fled. Waterton chased him all the way back to camp, "shouting to increase his fear." Characteristically, Waterton immediately sat down and penned a mock heroic account of his slugfest with the serpent in Latin hexameters. Waterton's son Edmund found the poem among his father's papers after his death.

Once, in Leeds, England, medical professionals invited Waterton to an evening séance comparing the effects of curare poison, which Waterton had collected on his first journey up the Demerara River, with the venom of some rattlesnakes an American was touring. But when the experiments were about to commence, no one—not even the American—would handle the rattlers. Seeing an opportunity for the kind of freakish showmanship he loved, Waterton stepped forward and volunteered to handle the snakes.

Reviewers of the *Wanderings* and naturalists in general had doubted Waterton's serpentine stories, and this public demonstration gave the Squire the perfect opportunity to set the record straight. Charles Waterton was not inventing his snake stories, though he may have, like any good storyteller, exaggerated. At Leeds he slowly put his hand into the glass container; holding one rattler after another firmly behind the head, where they couldn't reach to bite him, he brought them out and held them in his hands, while they bit the guinea pig or pigeon.

Halfway through the show, however, when Waterton's snake-charming had already stolen it, an eight-footer suddenly made a dart for freedom and crawled halfway out of the glass cage. Panic seized the doctors. Everyone except Waterton's friend Dr. Hobson took to their heels and fled the room. Several not only fled down the stairs but went out into the street, without their hats. Meanwhile the Squire calmly grabbed the escapee by the middle and put it quietly back into the box, as if nothing particular had happened. From that night, many more were convinced that Waterton's jungle stories were true.

There was a lot of devilment in Waterton, which did

not mix well with his alternate craving to be taken seriously. To the modern reader of *Wanderings in South America*, Waterton often seems to be a character in an early, not-so-accomplished novel by Laurence Sterne—Tristram Shandy's little brother, an addled, digressive, mock-pompous narrator, wandering around in a state of righteous befuddlement: a kind of barefoot Mr. McGoo, traipsing through the jungle. Indeed, if Waterton had not himself created Waterton the literary character, some novelist would surely have done so. He was simply too rich a character not to have been invented by someone.

Charles Waterton was born in 1782 in Yorkshire, England, scion of an ancient, untitled, aristocratic Catholic family whose lands and income had been greatly reduced during and after the Reformation. His childhood was spent at Walton Hall, which he inherited after his father's death and which became his home until his own death in 1865, at age eighty-three, when his foot caught for the umpteenth and final time in a bramble and he toppled headlong onto a fallen log, suffering fatal internal injuries. (While lying on the ground dying, he gave instructions to a laborer who happened to be with him to cut down several trees nearby to improve the view.)

From a very early age, we are told, Waterton was fascinated with nature and especially absorbed in observing birds. His earliest memory, recorded in his autobiography, was of climbing to the roof of the family manor house at age eight to get a closer look at a starling nest. He also displayed an extreme athleticism, such that his friends and biographers believed he was completely double-jointed: well into his seventies the Squire would entertain people by scratching the back of his head with his big toe.

Remanded by his father to the hands of the English

Jesuits at Stonyhurst School, young Waterton immediately showed himself a capable student but an incorrigible dodger. He skipped classes to climb trees. He fled to the woods to go hunting. On a dare he once jumped on the back of a bull, only to get tossed over the horns, perhaps previewing the cayman ride decades later.

Instead of punishing him, however, the merciful Jesuits seem to have recognized that Waterton's rule-breaking and misbehavior were more a matter of physical exuberance and love of creation than maliciousness. They sensibly managed to take advantage of the boy's natural inclinations by making him the school's ex officio rat-catcher, fox-trapper, rook-shooter, and leather football maker, the stuffing of which may have been among his first attempts at a type of taxidermy.

If Waterton may be said to have later become the *puer eternis* of tropical exploration, the Stonyhurst Jesuits can be thanked—or blamed. When he left Stonyhurst at eighteen, one of the priests warned him that inasmuch as his wild temperament would probably carry him to a roving life, he would be well advised to forgo alcoholic drink. Waterton not only swore never to drink, he kept that promise. Since rum was the regulation ruin of nearly all Europeans who went to the tropics, Waterton's teetotal may have had some influence on his extraordinary ability to withstand recurrent attacks of such diseases as malaria, typhus, and yellow fever, later found to be spread by mosquitoes, but then attributed to unhealthy vapors called miasma, said to be emitted in hot and humid environments.

Waterton received a regular but modest income from the family lands all his life, which freed him of the need to earn a living. When his school days were over, his father asked him to give up rat-catching in favor of the gentlemanly pursuit of riding to hounds. The young Squire rode so recklessly and dangerously, jumping blind fences in front of ditches and going at full gallop over rough-ditched terrain, that after a year his father begged him to quit. Waterton promised, and gladly went back to climbing trees.

After a first foreign trip to Spain, Malaga, and Malta, where he barely escaped a plague that killed his uncle and brother, Waterton found that he longed to bask his chilly frame in a hot sun. In November 1804 he went out to superintend the family sugar plantations in Demerara, which became the British colony of British Guiana. In those days, as well as today, the three Guianas—British, Dutch, and French—located on the northern coast of South America, were little more than huge sugar and spice plantations along the Caribbean coasts. Inland twenty-five miles stretched unbroken and unexplored forest all the way to the Amazon River.

The equatorial geography of these sugar colonies was much different from that of the American Gulf Coast and Mississippi Valley. As a result, African slaves more easily escaped into the forests and established communities of "maroons." Some attacked the white colonies but more intermarried with Amerindians and reestablished their African village cultures in the bush, where they remain to this day. The Guianas are still among the most remote and unknown parts of the Americas. How many can even name their capitals (Georgetown, Parimaribo, Cayenne)?

During his eight years administering his family's Demerara estates, Waterton was selected to carry diplomatic dispatches to the Spanish government in Orinoco. It was this trip up the Orinoco River that turned him into a jungle adventurer. Nothing could have been more thrilling for a young ornithologist than to feast his eyes and ears on the incredible variety of tropical bird life. The lowlands teemed with curlews and egrets, sandpipers and plovers, spoonbills and flamingos, frigate birds and pelicans; the uplands with parrots and toucans, cotingas and aras, finches and thrushes, houtous and jacamars. He heard for the first time the squawks of the horned screamer, a large gooselike bird with primitive hooks on its wings for climbing trees. He watched how the common vultures never feed till the king vulture is done. And he noticed the brilliant hummingbirds, feeding not only on flower nectar but also on the insects found there. There was no end to the fascinating bird life. "The naturalist may exclaim," Waterton said in the *Wanderings*, "that nature has not known where to stop in forming new species, and painting her requisite shades."

His dugout canoe was gliding along the Orinoco River one day when Waterton noticed a large "labarri" snake coiled in a bush along the river bank close by. This was the highly venomous fer-de-lance. Waterton shot but only wounded it. As the boat approached the bush he grabbed for the snake, intending to seize it by the throat and bring it on board. But the Spaniard at the tiller, realizing the lunatic *Ingles* was about to bring a live serpent into the boat, lost his head and veered hard for the deep water. Waterton just had time to grab onto the bush, where he remained dangling over the alligator-infested river, with the snake right above his head. Eventually he and the eight-foot fer-de-lance both got into the canoe. It was surviving this high-adrenalin incident that convinced Waterton to throw over plantation life and become an explorer.

"In the month of April, 1812, I left the town of Stabroek to travel through the wilds of Demerara and Essequibo," Waterton begins his remarkable narrative. Over the next thirteen years he made four expeditions into the interior of the Guianas, interspersed with periods of time living at an abandoned plantation on Mibiri Creek, near the present Georgetown. Here he must have witnessed new jungle retaking cleared land; such edge habitats are especially rich in flora and fauna as species invade and investigate the real estate.

Waterton was an outstanding field naturalist. There is little doubt he could enter into the minds of animals to fathom their intentions and understand their adaptations from excellent firsthand knowledge of their actual behavior. The *Wanderings* contain some of the most illuminating passages extant on the prolific wildlife of the American tropics, combining scientific accuracy and originality of observation with often lyrical expression—and supplying a great deal of what was then new knowledge.

We are now so used to zoos, museums, the *National Geographic*, and the Discovery channel, we scarcely remember that until explorers like Waterton described them, science knew very little about tropical wildlife. Nearly all knowledge in Europe and the United States came from dead and stuffed specimens. But as naturalists may attest, there is no such thing as a dead animal. For once life is gone, what remains is not the essential animal but merely its form in skin and bones. And forms can mislead.

The same may be said for captured specimens. Those who had written about the sloth, for example, Waterton said, "have remarked that he is in a perpetual state of pain, that he is proverbially slow in his movements, that he is a prisoner in space. . . . This is not the case."

In dead or captured sloths, observers thought they saw a quadruped with such weak, corkscrew back legs that it could not support itself comfortably on the ground as other quadrupeds did. They saw a pathetically maladroit creature with no soles on its feet and hooked claws. The name sloth itself had been given the animal in view of the incredibly awkward, tortuously slow manner in which captured specimens moved forward, often whimpering as they went. Since the word sloth also has religious connotations of sin and laziness, it was inevitable that the sloth should come to be widely viewed as an animal cursed by God.

"If the naturalists who have written the history of the sloth had gone into the wilds, in order to examine his haunts and economy," Waterton scolded, they would have learned that "the sloth's history must be written while he is in the tree." In the dark and gloomy forests, where few humans go, the animal is actually wonderfully well suited. "The sloth, in its wild state, spends it whole life in trees," Waterton continued, "and never leaves them but through force or accident. . . . And what is more extraordinary, not upon the branches, like the squirrel and the monkey, but under them. He moves suspended from the branch, he rests suspended from it, and he sleeps suspended from it. To enable him to do this, he must have a very different formation from that of any other quadruped."

Observed *in vivo*, the sloth "first seizes the branch with one arm, and then with the other; and after that, brings up both his legs, one by one, to the same branch; so that all four are in a line: he seems perfectly at rest in this position."

And so he is. Perfectly at rest and perfectly well adapted to an arboreal existence, feeding exclusively on leaves. The sloth's thick hair, Waterton noticed, is "so much the hue of the moss which grows on the branches of the trees, that it is very difficult to make him out when he is at rest." It was later learned that this camouflage, first described by Waterton, results from the sloth's fur actually playing host to mosses, an interesting example of adaptive coevolution.

Also receiving Waterton's attention were goatsuckers, or nightjars, a family of nocturnal birds, best known of which is the whippoorwill, whose insistent call fills soft summer nights. The name goatsucker derives from the bird's association with herds; it is often seen hopping up and down under goats or cows, and at least since the time of ancient Greece was believed by such activity to be stealing milk. "Father has handed down to son, and author to author that this nocturnal thief subsists by milking the flocks," said Waterton. For this reason, men had long viewed goatsuckers as pests and destroyed them on sight.

Waterton counted nine species of goatsuckers in his tropical jungles. By observing their behavior for himself, he concluded that, far from stealing milk, goatsuckers actually performed a useful function. "See how the nocturnal flies are tormenting the herd," he wrote, "and with what dexterity he [the goatsucker] springs up and catches them, as fast as they alight on the belly, legs, and udder of the animals. Observe how quiet they stand, and how sensible they seem of his good offices, for they neither strike at him, nor hit him with their tail, nor tread on him, nor try to drive him away as an uncivil intruder. Were you to dissect him, and inspect his stomach, you would find no milk there. It is full of the flies which have been annoying the herd."

The ant-bear, or anteater as we now call it, was another tropical species on which Waterton set the record straight. Waterton observed three species in the tropical forests. "The smallest is not much larger than a rat," he said. "The next is nearly the size of a fox; and the third is a stout and powerful animal, measuring about six feet from the snout to the end of the tail. He is the most inoffensive of all animals, and never injures the property of man. He is chiefly found in the inmost recesses of the forest, and seems partial to the low and swampy parts near creeks, where the troely-tree grows. There he goes up and down in quest of ants, of which there is never the least scarcity; so that he soon obtains a sufficient supply of food, with very little trouble."

Waterton described how this inoffensive animal, lacking teeth or swiftness for flight, nonetheless defended itself effectively from jungle predators. "Nature has formed his fore-legs wonderfully thick, and strong, and muscular, and armed his feet with three tremendous sharp and crooked claws. Whenever he seizes an animal with these formidable weapons, he hugs it close to his body, and keeps it there till it dies through pressure, or through want of food."

In European collections the ant-bear was always posed like other quadrupeds, with the foreclaws in the forward position like those of a dog. This was unreal, Waterton noted, for "the length and curve of his claws cannot admit of such a position." Instead Waterton had observed the ant-bear walking "entirely on the outer side of his fore-feet, which are quite bent inwards, the claws collected into a point, and going under the foot. In this position he is quite at ease."

The Squire, who more than once mentioned enjoying boiled ant-bear with his Indian acquaintances, likewise had the opportunity to discover another singularity of the animal's anatomy: "He has two very large glands situated below the root of the tongue. From these is emitted a glutinous liquid, with which his long tongue is lubricated when he puts it into the ants' nest. The secretion from them, when wet, is very clammy and adhesive," allowing the ants to stick to it.

Time and again Charles Waterton delights readers with his vibrant descriptions of tropical birds, snakes, mammals, insects. The glittering hummingbird darts through the air "quick as thought," fluttering from flower to flower to "sip the silver dew . . . now a ruby, now a topaz, now an emerald—now all burnished gold." The bushmaster snake, "unrivalled in his display of every lovely colour of the rainbow and unmatched in the effects of his deadly poison . . . glides undaunted on, sole monarch of these forests." The bird known as the tinamou "sends forth one long and plaintive whistle from the depths of the forest and then stops; whilst the yelping of the toucan, and the shrill voice of the Pi-pi-yo, is heard during the interval."

These passages are Charles Waterton at his best, describing a tropical nature that still sighs romance and power. It is in these amazing wildlife sketches, drawn straight from his own experiences, that Waterton gave us much that readers of the *Wanderings* can still enjoy today: the smell of tropical heat, the damp feel of the jungle floor, the gloom of the high canopy forest, the refreshing trade winds.

Waterton published three books during his lifetime—the *Wanderings*; a collection of natural history and personal essays; and an autobiography. *Wanderings in South America*, his first book, was written in a suspicious hurry after his fourth and final journey to the Guianas, an obscure trip whose motive Waterton failed to inform readers about, and whose duration he also neglected to mention.

After the shocking and sad death of his eighteen-year-old wife in childbirth in 1830, the Squire had lapsed into a period of grief and duty to his infant son. Then the reading of Alexander Wilson's *Ornithology of the United States* fanned the dying embers of adventure in him, and he set out for the United States with the stated intention of traveling west into the wilds of "bugs, bears, brutes, and buffaloes," as he put it.

Instead, as was often the case, he veered off course and screeched to a halt in Philadelphia. There he heaped lavish praise on the City of Brotherly Love, except for the streets "laid out in a grid pattern," which he found tiresome. In Philadelphia Waterton was well received as a naturalist and made influential contacts. Charles Willson Peale, the portraitist of George Washington, painted his picture.

At the time Waterton visited Philadelphia, Alexander Wilson had been dead only a few years. A close circle of Wilson's associates still lived in the city, both in the high society of the natural sciences and in the lower one of the printing business, where Wilson's *Ornithology* had been set in type. At the same time the ambitious John James Audubon was trying to succeed to Wilson's ornithological eminence; but Wilson's "family" in Philadelphia were putting up resistance.

From a delicate comment in Journey IV, we can be certain that Waterton wasted no time enlisting against Audubon. Here he is talking about Philadelphia: "From the press of this city came Wilson's famous 'Ornithology.' By observing the birds in their native haunts [i.e., just as Waterton had done], he has been enabled to purge their history of numberless absurdities, which inexperienced theorists had introduced into it." Audubon was clearly that "inexperienced theorist" introducing "numberless absurdities"—an unfair attack, since Audubon was just as much a leader in portraying animals from life and just as dedicated as Waterton or Wilson to observing wildlife "in their native haunts."

The Waterton-Audubon war of words continued for years afterward, focusing, incredibly enough, on the question of whether vultures identify carrion by scent or by sight. Those who, like Waterton, held that vultures "smell the effluvia" of rotting flesh as they fly, became known as "Nosarians." Audubon, who carried out experiments demonstrating that sight was the vulture's means of identifying prey, led the "Anti-Nosarians." No less a figure than Charles Darwin dipped his toe into the controversy when he tested the sight and smell mechanisms of a pair of condors captured during his voyage on the *Beagle*. He found no conclusive evidence one way or the other.

With only the skimpiest explanation—the Philadelphia mornings and evenings, Waterton complained, were too cold—the Squire left the United States by ship, and after touching Antigua, Martinique, Guadeloupe, Dominica, and Barbados in the Caribbean, he landed once more at Demerara. Wasting no time, he headed straight for the interior. But then, after only a few weeks, he returned to England, now with the excuse that "the rainy season was coming on," though it was December 1824 and the rainy season did not start until April or May.

What was the rush? Why did the Squire break off his trip to the United States, dash to Demerara, then bolt back to Britain?

"I mentioned in a former adventure," Waterton reminds us in his Journey IV discussion of howler monkeys, "that I had hit upon an entirely new plan for making the skins of quadrupeds retain their exact form and feature. Intense application to the subject has since that period enabled me to shorten the process, and hit the character of an animal to a very great nicety."

He had, on this odd trip, "procured an animal which has caused not a little speculation and astonishment. In my opinion, his thick coat of hair, and great length of tail, put his species out of all question; but then his face and head cause the inspector to pause for a moment, before he ventures to pronounce his opinion of the classification. He was a large animal, and as I was pressed for daylight, and moreover, felt no inclination to have the whole weight of his body upon my back, I contended myself with his head and shoulders, which I cut off: and have brought them back with me to Europe."

Waterton did not actually describe this animal, other than to say its features were "quite of the Grecian cast," whatever that may mean. After a page of coy speculation, including no less than four quotes from classical Latin—this from a man who absolutely refused to use Latin scientific nomenclature, which he called "jawbreakers"—the Squire wrote, "Now if we argue, that this head in question has had all its original features destroyed, and a set of new ones given to it, by what means has this hitherto unheard-of change been effected? Nobody in our museums has as yet been able to restore the natural features of stuffed animals."

The mystery creature was Waterton's infamous "Nondescript"—the head of a howler monkey he stuffed with such an artful touch and advanced taxidermy that it came out looking like an extremely hairy gentleman of the Georgian period. Somehow the impulse must have struck him in Philadelphia that if he went to the Guianas and obtained a howler monkey head, he could apply the new methods of taxidermy he had been working on for years, and pass the result off as a new species—what's more, the missing link between humans and apes. Could his warm reception by Alexander Wilson's circle in Philadelphia perhaps have encouraged Waterton to try to pull off this scientific hoax?

Waterton was proud of his advancements in the science of taxidermy. In an epilogue to the *Wanderings* he laid out his whole method of preserving birds. But he was also an irrepressible joker. He flawlessly sewed the upper half of a stuffed bird onto the lower half of a baby alligator, and shaped the faces of stuffed dogs and cats—and pigs—to look like famous Protestant church figures and British politicians. There was no harm meant in Waterton's japes, jokes, and hoaxes. It was simply a good prank to stuff a howler monkey and fob it off as the missing link, so greatly debated by sourpuss academics and churchmen at that time. But unfortunately for him, who so yearned to be taken seriously as a naturalist, Waterton had struck upon the very strategy certain to ensure that his wonderful *Wanderings* would be panned up and down England as the literary work of a freak, a liar, and, worst of all, an eccentric. It was always that last epithet that stung Charles Waterton most.

On his return home to Walton Hall after Journey II, the Squire decided to turn his estate into the world's first bird sanctuary. Waterton's travels into the jungles of the Neotropics had been regularly accompanied by the blast of his fowling piece. He was an adamant hunter, collector, and stuffer of the creatures he loved, and proudly tallied up the kill after each journey. He collected more than 200 bird specimens in Demerara during Journey II. On Journey III he says he collected 230 birds, 2 land tortoises, 5 armadillos, 2 large serpents, a sloth, and an ant-bear, as well as the infamous bucking cayman.

Yet the more he learned of birds' intimate habits, the more convinced he became that they deserved human admiration more than ammunition. In a particularly rococo section of Journey II, Waterton adopts the voice of the woodpecker to make a speech to Man, showing the progress of Watertonian, if not woodpecker, thought:

"Mighty lord of the woods . . . why do you wrongfully accuse me? why do you hunt me up and down to death for an imaginary offence? I have never spoiled a leaf of your property, much less your wood. Your merciless shot strikes me, at the very time I am doing you a service. . . . If there be that spark of feeling in your breast which they say man possesses, or ought to possess, above all other animals, do a poor injured creature a little kindness, and watch me in your woods for only one day. I never wound your healthy trees. I should perish for want in the attempt. The sound bark would easily resist the force of my bill: and were I even to pierce through it, there would be nothing inside that I could fancy, or my stomach digest. . . . Wood and bark are not my food. I live entirely upon the insects which have already formed a lodgement in the distempered tree."

Through field observation the Squire had discovered that the rationale for shooting woodpeckers was bogus. In a larger sense he was becoming aware of the needless and cruel destruction of all kinds of birds as part of the unreflective human habit, inherited from distant times, of waging war on nature. In 1817 Waterton decided that no killing of any bird—even marauding owls and hawks—would henceforth be allowed at Walton Hall. Until then, individuals had protected their private menageries, and lands had been preserved for hunting. But no one before Charles Waterton had ever conceived the far-fetched idea of simply protecting native and wild creatures. Inasmuch as Squire Waterton was the very first, some have suggested that all the Audubon societies that exist for the protection of wildlife would be better named Waterton societies—the Squire's archrival never having so much as disavowed gunning, let alone created a bird sanctuary.

Waterton had a high wall built around all 259 acres of his property. It ran for some three miles, eight to sixteen feet high, to keep out foxes, badgers, and especially human poachers. He next ordered his retainers that no guns were to be discharged on his property, and no dogs were to run loose. No boats were launched on the lake so as not to disturb the waterbirds. Waterton prepared various landscape plantings to feed and attract birds, and even set out shelters to encourage birds to breed on the property. Finally he erected small hiding places where he could observe birds, which he referred to as "substantial hovels." This was another original Waterton dodge. The substantial hovel later became known as the "hide" in England, the "blind" in America. Most scientific ornithological observation today is carried on from within such hovels. All this took many years to accomplish. But in the end, Waterton had loosed a genuine innovation upon the world, which would grow into the most valuable concept in the field of wildlife conservation—the wildlife refuge.

Throughout his long life Squire Waterton regarded

nature as his prime field of endeavor. He was too undisciplined for a literary man, too wild for a scientist. He was certainly no doctor, though his magnificent physical stamina survived all those bleedings and purges he inflicted on himself. But his medical curiosity, as opposed to his practice, knew no bounds. Waterton explored a thousand miles of jungle to bring back curare, which he hypothesized would cure rabies and hydrophobia. He was wrong, as it turned out, but that should not take away from his right idea that tropical wildlands contain thousands of natural productions unknown outside the tropics, and potentially beneficial. "What an immense range of forest is there," he wrote. "No doubt, there is many a balsam and many a medicinal root yet to be discovered, and many a resin, gum, and oil yet unnoticed."

Whatever he was not, Charles Waterton was certainly an outstanding field naturalist and explorer who did more than anyone of his time to take the study of tropical nature out of the dry cabinets of Europe and make it a fresh, lively, and exciting undertaking. He was a virtuoso in the jungle who combined tremendous physical courage and toughness with close observation and veneration of nature. That his book *Wanderings* has gone through so many editions shows there is something continuously fascinating about Waterton. He was one of those incorrigible near geniuses who kept his schoolyard swagger and his schoolboy stubbornness into his adult years. In the year of his death at age eighty-three, we are told, he still was able to climb to the top of the tallest trees, to read his favorite Latin poet, Horace, and to watch his beloved birds. To borrow the words of praise Trollope had for British hunting parsons, Charles Waterton rode hard to the end.

3

Charles Darwin: A Few Questions in the Americas

CHARLES DARWIN WAS fifty years of age when he finally published his theory of evolution in *The Origin of Species*. It was an overnight best-seller, which swiftly became the most famous book in the entire history of science, so that by the time he was fifty-two or so, *everyone* thought it worthwhile to do Darwin's portrait. Although the modest Darwin did not enjoy sitting for painters and photographers, he became despite himself one of the best-known faces ever. Who hasn't seen a likeness of Charles Darwin? Who doesn't know that he was about one year younger than Methuselah, just a bit more severe than Moses? Who has never stared in mute awe before that massive dreadnaught of a head—it had to be mansion-sized to house those voluminous brains—bald, domed, hard, immutable as marble, as if the thousands of BTUs produced by molecular collisions given off during all that contemplation of "descent with modification" had finally burned off the hair roots from inside. Who hasn't understood the message of that gigantic and solid mesa of a forehead, thrusting itself forward, peculiarly like a contemplative caveman with a 240 IQ, as if the theorist of evolution's ontogeny personally had to recapitulate the phylogeny of his predecessors. Darwin's image was an artistic rendering of intellectual certainty, invulnerability to doubt, impossibility to error. It was, by God, a British head.

This icon of Darwin as the Victorian brain par excellence is now deeply entrenched in our culture of do-it-and-don't-look-back. Students of the natural sciences are especially prey to the equation Darwin = dull dehydrated facts. Even so accessible and fresh a document as *The Voyage of the Beagle*, which has gained popularity even as Darwin's theoretical works have slid into the slough of neglect, cannot change the suspicion of today's lip-readers that Darwin's life must have been a dreary exercise in crunching bits of information. On the contrary, Charles Darwin was young and lively, impressionable, and even foolish at one time—the very opposite of what you might expect from the future man of scientific genius.

Indeed, if anyone were ever seeking a traveling companion to the American tropics full of youthful enthusiasm, insatiable curiosity, good humor, the spirit of adventure, and a gentlemanly sense of honor, he would not have to look farther than young Charles Darwin as he came bounding up the gangway at Devonport, England, to board the *HMS Beagle* for her surveying voyage to South America. Darwin was all of twenty-two years old at that Christmastime 1831; a twenty-two-year-old Charles Darwin is itself a fact worth pondering.

On his arrival in the Americas at the end of February 1832, Darwin was a neophyte in the Neotropics. He was just down from university—first Edinburgh, where he had tried to study medicine in the footsteps of his doctor father and illustrious grandfather, Erasmus Darwin, but found he had literally no stomach for it—this was the time of medical procedures without painkillers; and then Christ's College, Cambridge, where his disappointed father had decided he should study to become a parson, but where the younger Darwin found horseback riding, partridge shooting, and carousing with his chums more to his liking. Young Darwin did earn his degree—just barely. As he admitted many years later in his autobiography, "When I left school . . . I was considered by all my masters and by my father as a very ordinary boy, rather below the common intelligence."

A bit of modesty never hurt. Darwin seems to have wasted precisely enough of his college days to temper his high spirits with a pinch of Victorian shame, which was a major motivation to "do something" with his education. He was the son of a wealthy physician, so there was no pressing need for him to earn money. His interest in nature and natural history seems to have come about partly as an extension of the boyish pleasure he took in the gentlemanly outdoor pursuits of riding and hunting. The great naturalists were all by temperament outdoors types—walkers, campers, mountaineers, hunters, fishers. Darwin did not differ. From boyhood he collected minerals, and during his schooldays he had read the English country parson Gilbert White's hallowed *Natural History of Selbourne*, a book of such magnetic charm in its exegesis of natural history field-tripping that it may be fairly called the Book of Genesis of nineteenth-century natural history literature.

Darwin had become, like many young naturalists of the time, an avid collector of natural history objects, from birds' eggs to pressed flowers, insects, butterflies, shells, and

HMS Beagle, which brought the twenty-two-year-old Charles Darwin to South America, the crucible of his theory of evolution.

so on. A famous story of the young collector also comes from the *Autobiography*: "I will give you proof of my zeal," he says. "One day, on tearing off some old bark, I saw two rare beetles, and seized one in each hand; then I saw a third and new kind, which I could not bear to lose, so that I popped the one which I held in my right hand into my mouth." The angry beetle squirted a burning chemical into the mouth of its human container, and Darwin had to spit it out.

Darwin's formal training in the sciences was meager. But as his uncle Josiah Wedgewood described him, he was a young man of "enlarged curiosity"—a brilliant turn of phrase which reduces an intellectual quality to a rather mechanical one, and treats curiosity as a kind of body organ. Darwin was also the child of a no-nonsense Protestant upbringing, the product of a high-achieving professional family. If not installed in his genes, the work ethic was clearly stamped all over his environment. Individuals are never separate from the cultural climate of their time, and this was the era of British industrial capitalism approaching its zenith. The idea of progress—in the sciences, in society, and in individual life—was ensconced at the top of the values heap, and the man of the hour was the bearer of that progress, the man of action and boldness, the outgoing man of the world —*homo faber*. Not a part of the opening decades of the nineteenth century was the affected world-weariness that would soon overtake the European upper classes in the form of romanticism; nor, perish the thought, the nihilism of the early twentieth century, nor the market Darwinism of our own consumer age.

This was an age of bright prospects of *carpe diem* in a rising British imperium. The internal, familial, and social pressures to apply himself were interlocking and overwhelming. And as these existed side by side with his personal condition of leisure and financial security, Darwin was in a rather good position to follow his naturalist's avocation on board the *Beagle* for what promised to be a five-year journey without pay.

In the first third of the nineteenth century the intellectual debate over creation was not solely or even mainly a biological debate but also a geological one. Mounting evidence from the fossils of extinct animals and plants discovered in geologic strata had already undermined orthodox creationist dogma, which asserted, first, that the earth and all its life forms were about six thousand years old and had been created by God in six days; second, that life forms were immutable, never changing from the way the deity had created them; and, by extension, third, that the extinction of species was impossible.

Then, in 1785, a Scottish physician, James Hutton, who had taken up geology as a hobby, published a book called *Theory of the Earth*. In it he described the manner in which the action of water, wind, weather, and fire (in the form of vulcanism) had slowly changed the earth's surface. Winds eroded exposed rock. Water, too, eroded the land, carried silt down rivers, and deposited layers of sediment in estuaries and on sea floors. Volcanoes threw lava, which cooled to form new rock. Hutton maintained that such natural forces had always proceeded in the same manner and at the same rate. This became known as the "uniformitarian principle," and belief in the regularity of natural agents in geologic change became known as "uniformitarianism." Hutton pointed out that to account for such vast changes as the building of mountains or the gouging out of river canyons, vast ages of time were required. Therefore the earth had to be not six thousand but many millions of years old.

At about the same time the Frenchman Georges Léopold Cuvier was studying the anatomy of fossils dug from different strata of rocks. Cuvier compared the anatomy of prehistoric fossils to the anatomy of currently extant species, and saw that the extinct species represented in fossils had similar structures to the various living phyla, or groups. For example, some fossils had backbones, some living animals had backbones. Some fossils had external shells, and so did some living forms. Cuvier also saw that the deeper and thus older a fossil was, the more it differed from the existing life forms it seemed related to. He could place some fossils in consecutive order by age, seeming to demonstrate gradual change. By means of comparative anatomy, Cuvier developed the prospect of a gradual evolution of life forms parallel to the uniformitarian evolution of inanimate matter.

The prospect—but not the proposition. Cuvier was a pious Lutheran who would not accept a uniformitarian view of life, though he did understand that the earth itself must be more ancient than the Bible prefigured. Instead Cuvier proposed, in effect, an intellectual compromise that would take into account both deism and his own pioneering work in paleontology. His notion was that God caused the earth to suffer periodic calamities or catastrophes, wiping out all life. After each ruination the Lord then created an entirely new set of species, distinct but not completely different from those that had existed before. The fossil record, so troubling to Christian believers, actually marked these catastrophes. Cuvier proposed that the modern forms of life on earth had arisen after the most recent catastrophe—the animals Noah took onto his ark during the biblical flood. "Catastrophism" was the last scientifically dignified deistic defense against mounting geological evidence for gradual, noncatastrophic, uniform change.

This is roughly where the debate stood at the start of the 1830s, when two important events occurred. First, a British freelance writer and adjunct geology professor named Charles Lyell published *Principles of Geology*, brilliantly marshaling the uniformitarian case. Second, Captain Fitz Roy of *HMS Beagle* brought young Charles Darwin on board as ship's naturalist. And Darwin brought Lyell and Humboldt to read during his five years before the mast.

Now that we know a little something of our traveling companion, his time and its science, let us enter the gates of the tropics together with Darwin and accompany him on his first walks in an American tropical forest. We can share with Charles Darwin some of his very first impressions of tropical life, in a forest outside the town of São Salvador, Bahia, Brazil, on February 29, 1832, and later near Rio de Janeiro.

"The day has passed delightfully," Darwin recorded. "Delight itself, however, is a weak term to express the feelings of a naturalist who, for the first time, has wandered by himself in a Brazilian forest. The elegance of the grasses, the novelty of the parasitical plants, the beauty of the flowers, the glossy green of the foliage, but above all the general luxuriance of the vegetation, filled me with admiration. . . . To a person fond of natural history," he continued, "such a day as this brings with it a deeper pleasure than he can ever hope to experience again."

A little later at Rio, Darwin continued, "In England, any person fond of natural history enjoys in his walks a great advantage, by always having something to attract his attention; but in these fertile climates, teeming with life, the attractions are so numerous that he is scarcely able to walk at all."

And again, "It is easy to specify the individual objects of admiration in these grand scenes; but it is not possible to give an adequate idea of the higher feelings of wonder, astonishment, and devotion, which fill and elevate the mind."

One's first steps into a tropical jungle or rain forest can be overwhelming, even intimidating. There was so much going on, Darwin scarcely knew where to look first. And everywhere he looked—I am quoting a passage from much later on—"the land is one great wild, untidy, luxuriant hothouse." The working naturalist's inclination to build a generalized picture of a habitat by observing a series of discrete facts seemed suddenly thwarted by the sheer profusion of tropical forest life. As Darwin later concluded, in one of his most beautiful passages, the tropical forest was much more than the sum of its parts: "It is a hopeless attempt to paint the general affect. Learned naturalists describe these scenes of the tropics by naming a multitude of objects, and mentioning some characteristic feature of each. To a learned traveller this possibly may communicate some definite ideas: but who else from seeing a plant in an herbarium can imagine its appearance when growing in its native soil? Who from seeing choice plants in a hothouse, can magnify some into the dimensions of forest trees, and crowd others into an entangled jungle? Who when examining in the cabinet of the entomologist the gay exotic butterflies, and singular cicadas, will associate with these lifeless objects the ceaseless harsh music of the latter, and the lazy flight of the former—the sure accompaniments of the still-glowing noonday of the tropics?"

Here is no doubt what Darwin saw:

Highest overhead, the loftiest trees of the jungle form what is now called the emergent layer, because they are tall enough to emerge above the forest canopy. These are not necessarily the oldest trees, nor the largest in diameter, but many are heavily buttressed in their first story, to support their awesome height. You can often see no more than the trunk of emergent layer trees. The greater number, as Darwin observed, are not more than three or four feet in circumference.

Just below the emergent trees is the main canopy—tall trees with spreading crowns interwoven to form a continuous cover of foliage shading the ground below. The main canopy trees are somewhere between 75 and 150 feet tall, and, as Darwin observed, many are not more than 3 or 4 feet in circumference, though again, many are buttressed or grow upon stilt roots.

Darwin was most impressed in the canopy level by the palms growing amidst the branching trees. Palms, he said, "never fail to give the scene an intertropical character." At this early point in his travels, perhaps Darwin could not recognize many other canopy trees, which can be discouragingly difficult to identify. Characteristically, tropical forests do not contain copses or groves of the same species of trees but rather individuals at widely spaced distances. You will not see ten spondia trees growing in the same area; you will see one spondia tree, and then, several miles away, another single spondia. At 100 to 150 feet up, canopy-level trees usually have somewhat similar leaves, and identification of a tree can be made only by (1) recognizing the bark, or something unique about its form; (2) knowing the bloom, if the tree happens to be in flower; (3) knowing the fruit or nut fallen underneath (but only if a sufficient quantity lie directly under one tree: a single specimen of fruit could have fallen from any of a dozen interwoven branches); or (4) climbing the tree.

The rain forest canopy, a habitat closed off to nineteenth-century observers with the exception of animal specimens shot from the ground, has turned out upon late-twentieth-century investigation by means of slings and mechanically operated suspension platforms, to be where the ecological action is in the tropical rain forest. Recent estimates say that as many as two-thirds of the species in the rain forest live at the canopy level. The canopy traps rain and blocks wind, so it is a prime habitat for many of the more delicate parasitic plants and epiphytes (air plants, which grow on other plants and have no connection to the ground, such as bromeliads and orchids, which attach themselves to the bark of tree branches and extend air roots down to absorb moisture). Darwin also saw in the canopy woody creepers and vines, especially members of the liana family—Tarzan's favorite plant—which often take root beside a tall young tree and rapidly twist up to the canopy before leafing out there. Essentially the vines use trees to hoist themselves up to canopy level; once there, they spread their small crown out, strangle the tree, take its place in the sun, and live quite contentedly wrapped around the dead trunk.

Darwin continued, "If the eye turned from the world of foliage above, to the ground beneath, it was attracted by the extreme elegance of the leaves of ferns and mimosae." We are now into the so-called understory, which is fairly open and sparse in tropical forest owing to the fact that only two to five percent of sunlight penetrates the canopy, except where a tree has died or fallen. The myth of "impenetrable" jungle undergrowth, by the way, came from the accounts of explorers traveling along tropical rivers, where thickets grow on the banks responding to the gaps in the canopy provided by the river. Hack through these narrow walls of vegetation and you find on the other side that the jungle floor is actually remarkably open and quite easy to bushwack through.

In addition to absorbing most of the sunlight, the main canopy also acts as a roof preventing free air circulation, so the humidity in the dark understory stays high, and the temperature doesn't vary much, giving the jungle its characteristic hot, sticky feel. The understory consists of an open, sparse level of young and slender trees, tree ferns and dwarf palms, at about fifteen to fifty feet. Then a shrub layer where a few small full-grown trees and a few true shrubs grow. Then what's called the field level, where tree seedlings and soft-stemmed herbs can be found. Some of these soft-stemmed plants—the arrowroot, ginger, acanthus, and spiderwort families, for example—dominate the understory as bushy plants whose leaves tend to be broad and long, probably to help them absorb the scant light. These are the showy tropical plants that have become familiar to us as imported houseplants. Many of them flower in spectacular fashion; bright flowers and daring inflorescences seem required in order to attract pollinators in the shady forest. Finally we reach the jungle floor itself, a mass of rotting leaf litter and decaying branches. Very few photosynthesizing plants. Many types of fungi, which draw their nutrition from the decaying matter. Beetlemania.

These levels or layers, of course, are not neat delineations, marked off like the yard lines on a football field. When you look around in a tropical forest you are likely at first to see a jungle, as did Darwin. The main point is that the tropical forest should be understood as a multilayered affair, though the actual floor plan may vary widely by latitude, elevation, and other geographic factors. The classic evergreen rain forest in fact accounts for only about one-quarter of all tropical forests. There are also gradations of wet forest, moist forest, montane, premontane, and lowland, as well as true deciduous tropical dry forest where the trees drop their leaves during the dry season.

Whatever type of tropical forest one travels into, the exuberance of the vegetation is expressed in multiple layers. It is these several levels that provide the rich diversity of habitat for the animals. Many of the tropical forest's animals specialize in one type of plant in one specific layer of the jungle. This specialized interdependence between plants and animals often accounts for the many intricate and entertaining natural history stories to be found in tropical forests. These curious tales of interdependence, coevolution, harmony and warfare, altruism and sometimes criminality or fraud, are the reward for a naturalist's interest in the tropics.

One such story can be told about Charles Waterton's friend, the three-toed sloth. The sloth, as we know, is legendary for its slow movements. Three-toed sloths—*Bradypus variegatus*—are arboreal mammals that live, feed, and reproduce many meters above the forest floor, near the upper levels of the forest canopy. They feed almost entirely on leaves, using a large ruminantlike (cowlike) stomach and long intestinal tract to aid in the digestion of this energy-rich but relatively indigestible foodstuff. To obtain a mix in their diet, sloths every so often move to the crown of another tree, using pathways formed by lianas, which interlace the canopy. To climb down to the ground each time they wished to change trees would be wasteful of sloths' energy. And, as slow and defenseless as they are, sloths on the ground are extremely vulnerable to predators.

But there is one significant exception to the sloth's avoidance of terra firma. About once a week sloths climb down to the forest floor, dig out a depression underneath their tree with their stubby tail, defecate into the hole, urinate over it, cover over this week-long accumulation of waste products with leaves using their hind legs, and then climb back up to the canopy. The entire process of descending, defecating, and returning to the canopy usually takes about thirty minutes. During this period sloths are exposed to terrestrial predators, mammalian predators that hunt the trees, and even large avian predators such as harpy eagles. So the forces that promote this behavior must be strong.

Why do sloths do this? The only tenable hypothesis advanced so far is that over the ages sloths have developed an interdependent relationship with the individual tree they depend on for food. The slow decomposition of sloth feces promotes the tree's growth, helping to provide the sloth with a more constant and higher-quality food supply. In other words, the sloth is fertilizing his own tree, so that he can climb back up and eat the leaves that grow as a result of his weekly organic gardening efforts below.

The natural history stories that come out of tropical forests open up another world to us. Darwin understood that the multilayered forest was the framework for the tropic's biological exoticism when he wrote, later in the voyage: "How great would be the desire in every admirer of nature to behold, if such were possible, the scenery of another planet! Yet to every person in Europe, it may be truly said, that at a distance of only a few degrees from his native soil, the glories of another world are opened to him."

Of course, to Americans, those glories are even closer.

As they first approached the New World, near the outset of the *Beagle*'s surveying voyage, Darwin noted how the entire two-thousand-mile coastline of Brazil is formed of granite in various states of erosion. He asked, "Can we believe that any power, acting for a time short of infinity, could have denuded the granite over so many thousand square leagues?"

At the conclusion of *The Voyage of the Beagle*, commenting on the changes in the surface rocks of Ascension Island, Darwin asked philosophically, "Where on the face of the earth can we find a spot, on which close investigation will not discover signs of that endless cycle of change, to which this earth has been, is, and will be subjected?"

These two questions, posed at opposite ends of his journey, are really the same question. They bracket the sea change in his thinking that Darwin apparently underwent during his five years before the mast. They are old-fashioned rhetorical questions, not so subtle endorsements of the uniformitarian view: that the world of the past and the world of the present are one continuous world, developing over eons of time according to natural, regular, and gradual processes, such as erosion, sedimentation, subsidence, and uplift. In essence: the world we see before us is the product of an evolution of the inanimate world.

The Voyage of the Beagle, though written in journal form, was not a true journal. Darwin wrote the manuscript after his return to England, based on the facts and observations contained in the journals he had kept during the voyage but with numerous additions, clarifications, rectifications, and a little help from his scientific friends, including Sir Charles Lyell, author of *Principles of Geology*, the uniformitarian bible, which Darwin brought along to the Americas, read and reread during his travels, absorbed fully, and praised lavishly. In his book Lyell took up Hutton's uniformitarian cudgel against the catastrophists by closely reading the earth's layers and the fossils they contained. From England to Sicily, to the Canary Islands, to the Mississippi River Basin, Lyell marched on with his analysis of how natural forces had constructed the earth's surface, like an unchanging physical force himself. What are the transporting effects of currents? What part does running water play in the formation of deltas? How do earthquakes change the earth's crust? What is the relation between earthquakes and volcanoes? Phenomenon by phenomenon, place by place—almost rock by rock—Lyell hammered away with his geologist's pick, chiseling his marble principle, not only that everything on earth, inorganic as well as organic, could be explained and understood by natural forces operating over time, but that these forces had always acted the same way, and at the same rate.

Insofar as the final manuscript of the *Voyage* was a reconstruction, there is no neat answer to the question, "What did Darwin know, and when did he know it?" As a travel book with a subtext—his change of mind away from the diluvial defense and toward an evolutionary opening—*The Voyage of the Beagle* is a beautiful synthesis, a natural history picaresque where the points on the map of the American tropics he touched often became points of illumination. What can be said with certainty is that Darwin's experiences in the Americas were largely responsible for converting him from an uncertain Cuvierist to a rock-hard uniformitarian. Somewhere along the line he began to understand in his soul that evolution applies not only to geology, not only to the way the earth's surface developed, but to the way living things developed as well. Let us briefly trace Darwin's observations that seemed to have the greatest impact in transforming his thinking.

The *Beagle*'s first goal was to survey and map the east coast to South America, from Bahia, Brazil, southward to Rio de Janeiro, to Montevideo, Bahía Blanca, and thence to Tierra del Fuego at the southern tip of the continent, with a side trip to the Falkland/Malvinas Islands. During this phase of the expedition, Darwin left the ship several times for extensive land travels. He not only walked in his first jungle but rode a horse across the famed Argentine pampas and the dusty plains of Patagonia. There he became familiar with the marvelously varied animal life of Patagonia and wrote memorable descriptions of the rhea, the guanaco, the armadillo, the viscacha, the capybara, the agouti, and the equally wild gauchos, the cowboys of the Argentine.

But perhaps his most fantastic discoveries in Patagonia were the fossilized remains of gigantic prehistoric land mammals, which were then called megatheria but which we now call megafauna. These included bones and partial skeletons of the megalonyx, the scelidotherium, the toxodontia, a complete skeleton of the mylodon, and others—nine big quadrupeds in all, which Darwin recognized from their anatomy as extinct allies of modern elephants, rhinoceroses, armadillos, camels, horses, and sloths. Very little or nothing was known of the age, existence, or habits of any of these creatures at the time, other than the fact that since their bones lay close to the surface, they had been alive until relatively recently.

The simple structure of the mylodon's teeth, for example, indicated that these animals were gnawers, living on vegetable food, probably leaves and twigs. But their form was so ponderous, and their great curved claws so unsuited to locomotion, that some naturalists thought mylodons must have lived in the trees, like modern sloths. Darwin noted dryly that "it was a bold, not to say preposterous, idea to conceive even antediluvian trees, with branches strong enough to bear animals as large as elephants."

Later, back in England, Darwin brought these remains to his friend and colleague, the zoologist Sir Richard Owen, who offered an ingenious solution. Instead of climbing the trees, mylodons pulled branches down and tore up the smaller trees by the roots. "With their great tails and huge heels firmly fixed like a tripod on the ground," Owen suggested, the giant ground sloths used the colossal breadth and weight of their hindquarters like a fulcrum. Mylodons were also equipped with an extensile tongue like modern giraffes, to reach their leafy food. Thus what seemed to be impossibly clumsy, encumbered creatures actually turned out to be quite serviceable and well adapted for survival.

Darwin was inclined to observe the way species seemed so often to be modified in one way or another, enabling them to survive in a certain way. Thus in the Falklands/Malvinas he summarized three different bird species found in South America "which use their wings for other purposes besides flight: the penguin as fins, the streamer as paddles, and the ostrich as sails." Still, these were birds, and they possessed wings. Few would dispute the idea that God gave birds wings to fly. But if you believed that, then how to account for these alternative uses? Perhaps the birds were simply ignoring God. Seeing these modifications in species for particular purposes was the genesis of his challenge to the immutability of species. Darwin himself never called his theory evolution—it was "descent with modification." It was the modifications that so often fitted the animal for survival.

The discovery of the megafauna fossils on the dry plains of Patagonia excited Darwin to a rampaging herd of insights and speculations in his journals. They are the most passionate passages in the *Voyage*, perhaps the most interesting intellectually, an exceptional view into this restlessly questioning mind, applying itself full force to thorny and fundamental questions. For one thing, Darwin understood the continuity between the rich diversity of mammals currently on the sparse plains of the Argentine and the equally rich diversity of the extinct giant mammals. Having discovered half the skeleton of a *Macrauchenia patachonica*—an earlier pachyderm related to the rhinoceros and tapir—Darwin mused that "the relationship, though distant, between the Macrauchenia and the Guanaco, between the Toxodon and the Capybara—the closer relationship between the many extinct Edentata and the living sloths, anteaters, and armadillos, now so eminently characteristic of South American zoology . . . are most interesting facts.

"This wonderful relationship in the same continent between the dead and the living," he continued, speaking of whole species rather than of individuals, "will, I do not doubt, hereafter throw more light on the appearance of organic beings on our earth, and their disappearance from it, than any other class of facts."

A big, not to say bold, claim. But characteristic of Darwin's style, new facts were not only answers but the catalysts for formulating new questions. *The Voyage of the Beagle* is a megagame of "Twenty Questions," played out on jungle trails, over vast pampas, up and down the Andes, and through the eons of time. The fossilized bones of the extinct creatures raised in Darwin's mind the question, "What has exterminated so many species and whole genera?"

He starts testing answers. His first thought would be—some great catastrophe. But he looks at the geology of Patagonia and La Plata, and recognizes only that all the placid features of that land result from slow and gradual change. Could it have been a change of temperature, such as the glaciation of the Ice Age? But Lyell had reviewed all the evidence in *Principles of Geology* and determined that the megafauna thrived after the Ice Age. What about humans—could they have hunted the megafauna to extinction? Perhaps, but that would not explain the extinction of fossil mice and many other tiny creatures. What about drought? Darwin rejects that hypothesis: no one could imagine a drought severe enough, as he put it to "destroy every individual of every species from southern Patagonia to Behring's Strait."

Suddenly, as if a dam has broken, the questions come flooding out, and Darwin the young scientist is overtaken by a rhetorical fit: "What shall we say of the extinction of the horse? Did those plains fail of pasture which have since been overrun by thousands and hundreds of thousands of the descendants of the stock introduced by the Spaniards? Have the subsequently introduced species consumed the food of the great antecedent races? Can we believe that the Capybara has taken the food of the Toxodon, the Guanaco of the Macrauchenia, the existing small Edentata of their numerous gigantic prototypes?"

Calming sufficiently to reach a declarative sentence, Darwin concludes, "Certainly no fact in the long history of the world is so startling as the wide and repeated exterminations of its inhabitants." On the other hand, he continues (now at fever pitch), wherever we can trace the extinction of a species through the intervention of man, we know that "it becomes rarer and rarer, and is then lost." Darwin notes that

Lyell found evidence of this principle, too: species first become rare, then extinct. Today probably more people grasp this essential point about species extinction than any other fact of conservation biology: this idea constitutes our notion of endangered species. If an animal or plant becomes rare, watch out—it will probably become extinct. Stated in its elemental form: rarity precedes extinction. That is why so many biologists, seeing the massive number of species that have recently become rare, fear a wholesale extinction of species is coming. But a hundred years ago when Darwin wrote, extinction was still a mystery, not even considered a natural phenomenon by some. Cuvier and the catastrophists viewed it as a sudden act of God.

When Darwin was exploring La Plata and Patagonia, he had just come from the lush tropical forests of Brazil, where he had found very few large mammals. Confronted on the sparse plains with so many forms of quadrupeds, it started him thinking about southern Africa, the steppes and dry savannas, which also supported such large mammals as elephants, rhinoceroses, giraffes. Dry, sparse plains had supported the extinct herbivore megafauna in Patagonia. And this started him thinking about the mammoths discovered frozen and intact in Siberia. One supposition among scientists about the then-recent Siberian findings had been that only a vegetation of tropical luxuriance could have supported such behemoths. Because Siberia was not the least bit tropical, this theory had been used to support the idea that the climate had gone through sudden revolutions and catastrophic changes, leading to the congelation of a tropical creature in arctic tundra. With the evidence of the Patagonian megatheria before him, Darwin began to doubt out loud the catastrophist line.

But it was, finally, the massive evidence of the South American earth itself, coinciding with his study of Lyell's *Principles*, that pushed Darwin into evolutionary uniformitarianism. With plenty of opportunity to apply Lyell's principles of change to the scenes unfolding before his eyes—and giving way before his geologist's pick—Darwin became a crack geologist in South America. He absorbed that ultra-empiricism of analyzing strata of rocks, that puzzle-solving methodology, that satisfying logic, that architecture of the earth's past. Even those who consider the study of geologic formations a lifeless bore may still recognize the delightful mixture of precision and grandeur Darwin attained in descriptions like the following, of the great gravel and shell beds found for hundreds of miles along the Patagonian coast: "When we consider that all these pebbles, countless as the grains of sand in the desert, have been derived from the slow-falling of masses of rock on the old coast-lines and banks of rivers; and these fragments have been dashed into smaller pieces, and that each of them has since been slowly rolled, rounded, and far transported, the mind is stupefied in thinking over the long, absolutely necessary lapse, of years."

For the admirer of nature's works, the zest and literary craft that Darwin could bring to bear on descriptions is irresistible, like this one of the Maypu River in the Andean cordillera: "Amidst the din of rushing waters, the noise from the stones, as they rattled over one another, was most distinctly audible even from a distance. This rattling noise, night and day, may be heard along the whole course of the torrent. The sound spoke eloquently to the geologist; the thousands and thousands of stones, which striking against each other, made the one dull uniform sound, were all hurrying in one direction. It was like thinking on time, where the minute that glides past is irrecoverable. So was it with these stones; the ocean is their eternity, and each note of that wild music told of one more step towards their destiny."

Such vivid descriptions of the inanimate give a true sense of why the stones stirred Darwin's soul in South

America. For him, geology became the language of time, printed on the face of the earth. He learned to read these sandstones, conglomerates, granites, and mica-slates. He developed the ability to look at a given landscape and comprehend from the geologic facts (or words) what had taken place there and when: uplift, subsidence, erosion, slippage, injection. History was in every landscape. And when he climbed fourteen thousand feet into the Andes Mountains and found fossilized seashells there, he knew it meant that the highest elevations of South America had only recently, in a geologic sense, been at the bottom of the sea. "Daily it is forced home on the mind of the geologist," he wrote, "that nothing, not even the wind that blows, is so unstable as the level of the crust of this earth." For if such immense changes as the building of mountain chains from the seabed had taken place so recently in the crust of the globe, there was no further reason to believe in epochs of violent change.

By the time the *Beagle* turned away from the South American continent and headed west into the Pacific, Charles Darwin had abandoned catastrophic doctrines and in his heart and mind adopted the evolutionary view of change.

Until the *Beagle* arrived in the Galápagos Islands on September 17, 1835, Darwin had been looking at variation in species over the great vertical distance of time, and horizontally over very extensive geographic areas. Under such circumstances one was likely to invoke either geologic change or climate change, respectively, as the primary mechanism in evolution. In the Galápagos Islands we come to what Darwin later called "the great difficulty" of tortoise shells and bird beaks, a difficulty that was to catalyze his thinking about how evolution was accomplished.

The Galápagos archipelago consists of ten principal

islands, directly under the equator, five hundred to six hundred miles west of the Ecuadorian coast of South America. With his usual geologic zeal, Darwin noted that the islands were uniformly formed of granitic basalt of volcanic origin, some of the larger island craters rising to a height of four thousand feet. Contrary to what might be expected of islands on the equator, the tropical climate was moderated by the polar ocean current. The Galápagos lacked rain, but low-hanging clouds gave the upper parts of the islands, at an elevation of more than a thousand feet, a damp climate with quasi-tropical vegetation, while the lower parts were sterile.

Of the ten main islands, Darwin made a record of having visited five. The *Beagle* skirted from Chatham Island to Charles Island, around Albemarle and Narborough, and on to James Island. The distances between the islands as recorded by Darwin were as follows: Charles Island is fifty miles from the nearest part of Chatham Island and thirty-three miles from the nearest part of Albemarle Island. Chatham Island is sixty miles from the nearest part of James Island—but there are two intermediate islands between them not visited by Darwin. James Island is only ten miles from the nearest part of Albemarle Island, but the two spots where Darwin collected specimens on those islands were actually thirty-three miles apart.

The five weeks Darwin spent studying and collecting in the Galápagos archipelago eventually shook the intellectual world. Yet with an understatement typical of the man, Darwin claimed modestly in his journal that the natural history of the Galápagos was "eminently curious" and "well deserving attention." If he had any premonition that his findings in the Galápagos would start a chain reaction of thought whose inevitable conclusion would be the principle of natural selection as the mechanism of evolution, he revealed it in only the most subtle way.

The most notable inhabitants of the Galápagos were

extremely large land tortoises, which were hunted for food to supply passing ships. Darwin records his first encounter with the Galápagos tortoises unforgettably: "The day was glowing hot, and the scrambling over the rough surface and through the intricate thickets, was very fatiguing; but I was well repaid by the strange Cyclopean scene. As I was walking along I met two large tortoises, each of which must have weighed at least 200 pounds: one was eating a piece of cactus, and as I approached, it stared at me and slowly stalked away; the other gave a deep hiss, and drew in its head. These huge reptiles, surrounded by the black lava, the leafless shrubs, and large cacti, seemed to my fancy like some antediluvian animals."

Antediluvian: literally, before the deluge. What deluge? The one in the Bible that only Noah and the ark survived? Was Darwin here using the term in its conventional meaning of "before the biblical flood" to describe the antiquity with which these reptiles impressed him? Or did he have some inkling already of the controversy he would unleash with his theory of "descent by modification"? An eminently curious choice of word, that *antediluvian*.

The most intriguing fact about the Galápagos for Darwin was that a striking number of the flora and fauna were unique to those islands, found nowhere else in the world, though they showed a general relation to South American varieties. The 15 kinds of sea fish he procured, for instance, were all new species. Of the 16 land shells he found, 15 were peculiar to the Galápagos. Darwin collected examples of 193 flowering plants; 100 were new species confined to this island chain. Darwin procured 26 kinds of land birds; 25 were new species. With so many of its own special varieties, known nowhere else, the Galápagos were like a small world unto themselves.

The naturalist's excitement at discovering so many species new to science was equaled by his puzzlement at finding so many original creations in such a confined range. Darwin saw right away that neither geographic distance nor geologic time could account for all this variation. He mused, "Seeing every height crowned with its crater, and the boundaries of most of the lava streams still distinct, we are led to believe that within a period, geologically recent, the unbroken ocean was here spread out [i.e., that this was a relatively new area in geologic time]. Hence, both in space and time, we seem to be brought somewhat near to that great fact—that mystery of mysteries—the first appearance of new beings on this earth."

One example struck Darwin in a profound and lasting way. Of the twenty-five new bird species Darwin collected in the Galápagos, thirteen formed what Darwin called a "most singular group of finches." Not only were the finches unique to these islands, but all were "related to each other in the structure of their beaks, short tails, form of body, and plumage." Darwin classified the birds in the finch family Geospiza. They were all later named for him, becoming possibly the most famous little birds in the world—"Darwin's finches."

"The most curious fact," he wrote, "is the perfect gradation in the size of the beaks" in the thirteen different finches. Darwin shows that there are at least six species with minutely graduated beaks, and that in each of the thirteen the beak is slightly altered in size or shape such that the species could exploit some particular type of food. Some used their beaks to eat one type of seed, some another type, still others ate insects. "Seeing this gradation and diversity of structure in one small, intimately related group of birds," Darwin wrote, "one might really fancy that from an original paucity of birds in this archipelago, one species had been taken and modified for different ends."

Of the Galápagos tortoises—*Testudo nigra*—Darwin had much to say. On one level he was absolutely right in describing them as antediluvian, for the fossil record now shows that the Chelonian order of turtles and tortoises dates back to the Triassic period, some 200 million years ago. Chelonians are brilliantly designed to demonstrate that the best offense is a good defense, and a morphology like the external shell, which has lasted through the eons, must have a lot to recommend it. Nevertheless, I doubt there is another order of creature that has suffered at the hands of temporal irony like the turtles and tortoises. All that evolutionary energy that went into making Chelonians invulnerable to attack in the Age of Reptiles counted for little in the Age of Mammals.

Even in Darwin's day, though still numerous, the number of Galápagos tortoises had been greatly reduced by "hunting"—which hardly seems an apt word to describe the killing of the hard-shelled plodders. Darwin recounts that passing ships had been known to carry off as many as seven hundred tortoises. For hundreds of years seafarers had depended mainly upon tortoise and sea turtle for provisions at sea. This long-term decimation helps account for the precarious situation of sea turtles, in our own day, when all species are rare and in danger of extinction.

Darwin tells us what the tortoises eat (chiefly succulent cacti, leaves, an acidic berry, and lichen). He tells us that despite populating the arid lowlands, they are very fond of water (he found worn paths on the islands leading to springs and watering holes at the higher elevations), and he describes these huge creatures "travelling onwards with outstretched necks, and another set returning, after having drunk their fill. When the tortoise arrives at the spring, quite regardless of any spectator, he buries his head in the water above his eyes, and greedily swallows great mouthfuls, at the rate of about ten in a minute."

Darwin discusses the breeding season (the male utters a hoarse "roar or bellow," the female never uses her voice) and the laying of eggs (the female deposits them together in a hole in the sand, then covers them up). He gives the dimensions of the egg (a little larger than a hen's egg) and comments that the newly hatched tortoises fall prey in great numbers to buzzards, whereas the old tortoises seem to die only from accidents—falls down a precipice, for example—none ever having been found dead without some evident cause. He reports them as absolutely deaf (remarking how amusing it was "when overtaking one of these great monsters," that the instant he passed, "it would draw in its head and legs, and uttering a deep hiss fall to the ground with a heavy sound, as if struck dead"). Finally, Darwin tells us he frequently got on their backs and tried to ride them like Charles Waterton on his gator.

Only toward the end of his stay in the Galápagos did Darwin become aware of what he calls "the most remarkable feature in the natural history of this archipelago." In a conversation with the Galápagos Islands' Vice Governor Lawson, Darwin was told that tortoises from the different islands in the chain were different, and that the inhabitants could distinguish with certainty which island a tortoise came from. When he finally realized that each separate island had a varying tortoise peculiar to that island, you can hear Darwin slapping his forehead: "I never dreamed that islands about 50 or 60 miles apart, and most of them in sight of each other, formed of precisely the same rocks, placed under a quite similar climate, rising to a nearly equal height, would have been differently tenanted."

At this point Darwin backtracked through his collections and found that time after time, examples of species he had obtained thinking them endemic and aboriginal to the whole archipelago were actually, like the tortoises, specific to one island. Unfortunately, scrupulous as Darwin always was in marking his specimens, "most of the specimens of the finch tribe were mingled together." Not conceiving it possible that individual islands in the archipelago could have separate species, he had mistaken the islands as one entity, and the thirteen species of finches as coming from that entity. Now he had "strong reasons to suspect," as he put it in the journals, that some of the finch species were confined to separate islands. He goes on: "If the different islands have their representatives of Geospiza, it may help to explain the singularly large number of the species of this sub-group in this one small archipelago, and as a probable consequence of their numbers, the perfectly graduated series in the size of their beaks."

It was this new series of facts concerning variation in isolation that forced Darwin ultimately to abandon geologic change, geographic range, and climate as possible mechanisms for the modification of species. On his return to England he spent twenty years painstakingly searching for an answer to the question raised by his experience in the Galápagos: how could those finches, those tortoises, have become modified into different species on those small islands within sight of one another?

So Charles Darwin's travels came to an end. It was time to tote up the gains and losses of a circumnavigation of the world. For Darwin, who spent most of his career as a sailor suffering from seasickness, the boasted glories of the illimitable ocean were "a tedious waste" and "a desert of water." Darwin advised the potential adventurer to stay home unless he possessed "a decided taste for some branch of knowledge" to keep him going. Even the pleasure derived from admiring the scenery of various countries and beholding the different cultures of mankind could not offset the evils of isolation, hurry, and the loss of domestic society and family. He was strongly inclined to believe that the greatest reward of travel in the tropics awaited the individual who could go in the spirit of inquiry and harvest the satisfaction of knowledge and understanding. Since every tropical landscape was dominated by plants, Darwin concluded, "a traveller should be a botanist." Indeed, in the American tropics, a plantsman is never disappointed.

Darwin's travels in the American tropics convinced him that evolution was a fact. Although the ideas originated in Europe, it was South America that was the crucible of Darwin's theory. Twenty years later, when he had devised natural selection as evolution's operating principle, Darwin testified to the powerful impact of his treks through jungle and pampas. In the opening paragraph of *The Origin of Species* he paid homage to that land of fecund questions: "When on board HMS Beagle as naturalist, I was much struck with certain facts in the distribution of the organic beings inhabiting South America, and in the geological relations of the present to the past inhabitants of that continent. These facts . . . seemed to throw some light on the origin of species—that mystery of mysteries. . . ."

4

The Collector: Alfred Russel Wallace in the Amazon Valley

"IN ABOUT TWO DAYS MORE we were in the Amazon itself, and it was with emotions of admiration and awe that we gazed upon the stream of this mighty and far-famed river. . . . What a grand idea it was to think that we now saw the accumulated waters of a course of three thousand miles; that all the streams that for a length of 1,200 miles drained from the snow-clad Andes were here congregated in the wide extent of ochre-colored water spread out before us!"

So wrote Alfred Russel Wallace as he and his partner Henry Walter Bates entered the main channel of the Amazon River in August 1848. It was their brash idea to travel up the Amazon Valley, following the main stream and its tributaries, an area at that time, wrote Wallace, "as completely unexplored as the interior of Africa." To give you an idea of the magnitude of such an expedition, here are some preliminary facts to contemplate about the world's queen of rivers.

Item: Wallace fell short in his reckoning; modern calculations show the Amazon River actually flows some 3,900 miles from its sources on the eastern slopes of the Andes, which are less than 150 miles from the Pacific. In effect, the Amazon bisects the entire South American continent east to west. Of all the rivers on earth, only the Nile is longer (232 miles), but the Amazon makes it up in volume.

Item: The Amazon is not one river but the main artery of an enormous family of rivers. It is considered to have fifteen tributaries each more than one thousand miles long, and ten thousand tributaries in all. The British ecologist Norman Myers calculated that if you added the lengths of all of the Amazons together, it would wrap around the globe more than twice.

Item: For sheer volume of water, no other river comes close. The Amazon holds fourteen times as much as the Mississippi, seventeen times as much as the Nile. At its mouth the Amazon is more than one hundred miles across. The Amazon Delta is so large it contains an island called Marajó (*Ma-ra-zhu*), which has about the same area as Switzerland. Where the Amazon meets the ocean, it disgorges one-fifth of all the fresh water on earth—eleven times greater than the mighty Mississippi. It freshens the ocean water for hundreds of miles out to sea. This is how the Amazon was discovered in 1500, when a Spanish expedition sailing along the equator suddenly found itself in turbid fresh water, and followed it in.

Item: The true riches of the river are in the watershed it drains and the tropical rain forests that grow there. "The basin of the Amazon," wrote Wallace, "surpasses in dimen

The great collector, Alfred Russel Wallace, who spent four years along the Amazon river system.

sions that of any other in the world." He figured the area at 2.33 million square English miles. Landsat satellite images have helped modern geographers recalculate the Amazon basin at 2.7 million square miles. For the river's first thousand miles, where large tributaries like the Toncantins, Xingú, and Tapajós flow in from the south, Wallace reckoned the width of the Amazon forest at about four hundred miles north to south. Thereafter the jungle widens out to more than a thousand miles. Such an immense forested area led Wallace to proclaim the New World "pre-eminently the land of forests, contrasting strongly with the Old, where steppes and deserts are the most characteristic features."

When the Amazon forest suddenly began disappearing in the 1970s, scientists became alarmed about the consequences, not only for local ecosystems but for the global environment. The Amazon River Basin is the largest polygon of biomass on the planet, and it is plants, after all, that manufacture the oxygen we breathe. The rain forest's elimination would likely alter temperatures and adversely affect the hydrological cycle over a significant portion of the earth.

Amazon forests managed to avoid major trashing until the 1960s, when the nine-thousand-mile Transamazonian Highway was constructed to give access to logging, mining, and settlement. Little attention was given at the time to the environmental impact of the $500 million road, not to mention its effect on the numerous native tribes living there. It was not the result of evil developers plotting to destroy the planet's atmosphere so much as the Brazilian desire to "conquer the jungle" and the economic processes that conquest unleashed. Mistakes were made simply from lack of knowledge. Amazonia had been little studied—and never correctly mapped. Accurate maps depicting the river with its tributaries and wetlands were wanting. Understanding of the many ecological connections between land, river, climate, and development was absent. As the highway plunged westward, poor, land-hungry Brazilian pioneers followed, setting the jungle to the torch. Within a few years satellite images were available and showed spur roads proliferating, settlements growing, and forests shrinking. The jungle's topsoil is so thin that removal of the forest cover leads to swift erosion, even desertification. Proud planners responsible for the Transamazonian Highway originally estimated the completion of the "conquest"—i.e., the disappearance of the entire Amazon jungle—by the year 2050.

But the Amazon is not only so many billions of gallons of water and other impressive statistics. Amazonia is universally recognized as the tropical headquarters of life on earth. The Amazon rain forest not only symbolizes the vast diversity of living things but represents the power of plants through photosynthesis to manufacture the oxygen that fuels the earth's atmosphere and the water that gives our planet its distinctive green and blue colors. Amazonia is one of the lungs of the planet. The unconsidered devastation of the Amazon proved such an affront to universal consciousness that an international movement has sprung up to try to stop the destruction of the Amazonian rain forests before it is too late. (Wallace, by the way, was all for logging the Amazon forests. "The woodsman's axe has been the pioneer of civilization," he wrote. But Wallace also thought the supply of trees "inexhaustible" and the tropical soil "highly fertile." We know now that this is not the case.)

Before Wallace and Bates, the upper Amazon and Rio Negro, the main northern tributary on the upper branch, had never been explored by Europeans. The two Brits' plan was to travel up the Amazon Valley collecting animal specimens: insects, fish, birds, and mammals, principally, but in fact any kind of interesting specimen, fossil, mineral, fact, or artifact. Wallace and Bates were the greatest of scientific collectors, hard-bitten by the impudent little demon that makes people collect certain objects, whether art or rugs, books, bones, stamps, or baseball cards.

The collectors' demon first plants the desire to obtain the best of something, something no one else possesses—or, as in Wallace's case, a full set of something. The object for the natural history collector is a specimen of every species in a given category. Once the spark has been lit, the demon then proceeds to incite the hunt. No wonder Wallace's book, *Travels on the Amazon and Rio Negro*, often reads like a stiff-collared Victorian's hunting journal: "We remained here some days and had very good sport. Birds were tolerably plentiful, and I obtained [i.e., shot] a brown jacamar, a purple-headed parrot, and some fine pigeons."

Unlike collectors of abstract art or château wines, however, Wallace was not collecting objects associated with one's social status, but scientific facts. This is an important distinction. Wallace the collector wanted to find as many new species as he could, but Wallace the scientist collected facts/specimens in order to "solve the problem of the origin of the species," as he once put it in a letter to Bates. That is, how—by what processes—do new species emerge? Wallace's specialty was the geographic study of animal distribution, which would today be called biogeography or zoogeography. Wallace collected specimens to find out their variety and ranges—where and in what habitats they lived. The knowledge of how Amazonian species were adapted to their habitats, Wallace reasoned, could shed light on the way evolution works.

Wallace is often compared to Darwin because of the strange circumstance that the two nineteenth-century Englishmen, working independently of each other at the same time, both struck upon natural selection as the agent of evolution. Both had been to the American tropics, Darwin as naturalist on the *Beagle*, Wallace during four years collecting in the Amazon Valley. They shared paternity of the theory of evolution, but Wallace deferred to the older and more prominent Mr. Darwin, who received the fame as well as the blame.

But Wallace was no Darwin. He never tiptoed around biblical creation. He never harbored doubts that evolution was a fact: new species evolved according to natural processes; others became extinct. He noted in the margins of his personal copy of Swainson's *Treatise on Geography and Classification of Animals* (1835) that the whole effort "to reconcile science to scripture" was a waste of time for men of science. And Alfred Russel Wallace was preeminently a man of science. His whole trust was in natural laws, and he spent his career studying them in nature's great law libraries, like the Amazon Valley and the Malaysian archipelago.

Unlike Darwin, there was neither surgeon nor bishop nor barrister at the bar in Alfred Wallace's background. He was a working-class orphan in the industrial England of Dickens and Victoria, who described himself as a child as "shy, awkward, and unused to good company." Wallace had to quit school to go to work at Christmastime 1836; he was thirteen years old. He was apprenticed to his brother, a land surveyor, with whom he worked for the next seven years. In those days, surveyors walked. They walked to the job, walked on the job, and walked between jobs, from one to the next. Wallace walked all over the English moors and lakes and the Welsh hills, and it was during these long tromps as a surveyor's apprentice that he first became curious about nature. His start was modest: wouldn't it be good, the young Wallace thought, to know the names of the plants?

In 1841 he purchased an inexpensive book on botany so that he could identify the plants he encountered on his surveying travels. Thereafter he always carried it in his pocket. One book led to another. Soon Wallace was teaching himself Latin so that he could understand scientific nomenclature. He was a brilliant autodidact: everything he ever learned about biology, anthropology, or economics he either taught himself through reading or deduced from his own fieldwork. Nowadays people may think of a naturalist as someone who cavorts in the nude, but in the nineteenth century natural history was a popular and respected avocation for men of intellect and action, and one of the few pursuits in Britain that cut across class lines. In the democracy of science, a Mr. Nobody like Alfred Wallace could (and did) become the equal of a Charles Darwin.

Soon identification no longer satisfied Wallace; he was pricked by that obscure desire to collect natural specimens. In his autobiography, written many years later, Wallace tells of his growing devotion to his hobby with spiritual fervor: "Even when we were busy I had Sundays perfectly free, and used them to take long walks over the mountains with my collecting box, which I brought home full of treasures. . . . At such times I experienced the joy which every discovery of a new form of life gives to the lover of nature, almost equal to those raptures which I afterward felt at the capture of new butterflies on the Amazon." The collectors' demon had fingered him.

Preoccupied with his joys and raptures, Wallace left the surveying trade and found work as a teacher at a private school in Leicester. He spent most of his time in the school library reading Humboldt's *Travels* and Malthus's essays on population. By 1842 he had read Darwin's account of his voyage on the *Beagle*. Humboldt and Darwin lured Wallace to the American tropics.

It was also at Leicester that he found his friend and soulmate, Henry Walter Bates. Bates was another collector, apprenticed in his father's small hosiery factory, where Henry swept out in the morning and locked up at night. Compared with the humdrum of shifting bobbins in the mill, the collectors' demon offered Bates knowledge, beauty, and healthful outdoor exercise. He was way ahead of Wallace as a collector. While Wallace was still concentrating on plants, Bates had diversified into beetles and butterflies. Wallace was stunned "to find the great number and variety of beetles, their many strange forms and often beautiful markings or colouring" in Bates's collection.

It was like giving an addictive drug to an innocent child: Wallace was instantly hooked. He had soon bought the few necessary tools—stoppered bottles, tweezers, magnifying glass—and begun his own collection. From that time on, capturing critters became the chief passion of Alfred Wallace's existence.

Why limit collecting to England? Even though Wallace still considered himself a rank amateur—he could identify only a "moderate proportion" of his own insect collection—in 1847 he suggested to Bates that the two of them quit England and head out for the Amazon River to pursue their collective dreams.

They left Liverpool on April 26, 1848, and arrived at Pará (now Belém), the Brazilian port of entry for the Amazon. During their first two years on the Amazon they traveled and collected together, as far as the old town of Barra, now Manaus, at the convergence of the Amazon and Rio Negro. Here Wallace and Bates parted company, though neither man explains why in the accounts of their travels. Perhaps the two collectors wished to work separate territories; after all, the Amazon Valley was easily large enough for both of them.

Wallace found passage on Indian dugout canoes traveling up the Rio Negro and continued to explore and collect at a great rate for another two years. He did not try to identify every bug and butterfly while in the jungle. He would bring his collection back to England to classify the specimens, sell off the duplicates to pay his expenses, then study his facts and draw his conclusions. At least that was the plan.

A Narrative of Travels on the Amazon and Rio Negro, first published in the autumn of 1853, is Wallace's account of his expedition up the Amazon Valley. The first three hundred pages are a lengthy and dense (i.e., endless and boring) journal, recording Wallace's itinerary and travel arrangements, acquisitions, where he slept, what he ate, what time it rained, and how often his aboriginal guides yawned in their sleep. Wallace was no gifted storyteller. His sober concentration on the accumulation of facts makes *Travels on the Amazon* almost as tough going as the jungle itself. It may also distress modern readers to find the satisfaction Wallace took in shooting monkeys and birds out of the trees, but if we are to traverse the Amazon Valley with Alfred Wallace we had better get used to it: he collected ninety varieties of monkeys during his journey. The man was a prodigious hunter-gatherer.

The second part of *Travels on the Amazon*, however, is a short and splendid scientific description of the Amazon Valley, the first detailed account of its "physical peculiarities," as Wallace termed them. Wallace was one of the nineteenth century's most gifted natural scientists, at a time when a single individual could still become familiar with all scientific aspects. He wrote meticulously of the valley's physical geography and geology, its climate and vegetation, as well as the zoology and distribution of species. His book is the first attempt to describe the ecological system of the Amazon Valley as a whole, and anyone who is going there, or may go there, or dreams of going there some day, can still benefit from it.

Overall Wallace thought the most striking features of the river itself were its vast smooth waters (generally three to six miles wide); its "pale yellowish-olive" color; the great beds of aquatic grasses growing in abundance on its shores; the amazing quantity of branches and trunks and other forest matter it carried; and its level banks "clad with lofty unbroken forest."

From 4 degrees north latitude to 20 degrees south, Wallace said, the Amazon drained every river flowing down the east slopes of the Andes Mountains. Yet "our knowledge of the accuracy of most of the tributaries of the Amazon is very imperfect," he admitted, adding that they were "put in maps quite by guess." Speaking of guessing, Wallace, for example, named the Marañón River as the main source of the Amazon and reported that river's origin was in Lake Lauricocha, Peru. It was from Lauricocha to the Amazon's mouth that he (mis)calculated the river's length, "following the main curves, but disregarding the minuter windings." When satellite photography became available for the first time in the late 1960s, geographers set the source of the Amazon at the head of the Apurímac River, 225 miles south of Cuzco, Peru. They also discovered that many tributaries were completely mismapped, main curves missed, and "minuter wanderings" wholly absent, with islands as much as 60 miles across not even appearing.

Wallace calculated the river's flow at 500,000 cubic feet of water per second in the dry season. But at flood stage, he said, "it will amount to 1,500,000 cubic feet per second"—a tremendous lot of water. Wallace said the water volume should be understood relative to the area and climate of the watershed. The basin acts like a giant sponge. The equatorial forests covering it allow the heavy rains to penetrate, absorbing water but not giving it up through rapid evaporation, "as when [rains] fall on the scorched Llanos of the Orinoco or the treeless Pampas of La Plata." The superior absorbency of the tropical jungle accounts for its fabled luxuriance of flora. As Wallace put it: "Perhaps no country . . . contains such an amount of vegetable matter on its surface" as the valley of the Amazon.

Darwin said every traveler should be a botanist. In the

Amazon Valley, Wallace concurred. The "forest is the great feature," he wrote. The entire extent of the Amazon Valley was covered with "one dense and lofty primeval forest, the most extensive which exists upon earth." This forest, he observed, is distinguished by the "peculiar distribution" of its great variety of species of trees. In the Temperate Zone, as noted earlier, extensive tracts may be covered with a single kind of tree—oak, pine, or beech forests—but not so in the tropics. Of all the many species of trees, shrubs, vines, and herbs found in the Amazon, Wallace said, "we scarcely ever see two individuals of the same species together." Wallace thought this distribution pattern would probably prevent an intensive timber trade in the Amazon, but this was a hundred years before clear-cutting technology.

Wallace barely noticed the trade in rubber beginning to happen right around him. By far the most influential plant product of the Amazon was not mahogany or cocoa but rubber latex, which emerged from the tropical jungle to roll American society forward into the automobile age. Individual rubber trees, according to Wallace's rule of distribution, are separated by great distances in the jungle. The Brazilian rubber tappers were accustomed to such an itinerant harvest of the rubber latex. Then it was found in England that rubber trees brought from Brazil could be grown on plantations. Wallace belonged to the generation just before the big Amazonian rubber boom of 1870–1890, after the invention of vulcanization really put rubber on the road. He had only two locational sentences to say about the rubber tree: "the chief district from which india-rubber is procured is in the country between Pará and the Xingú. On the upper Amazon and the Rio Negro it is found, but is not yet collected." Pará rubber (*Hevea brasilensis*) from the lower Amazon came to be considered the finest commercial rubber in the world.

What about the Amazon's famous yellow-olive hue—

a color that has spelled "tropical jungle river" in a thousand Hollywood movies? All the waters of the river system could be divided into three basic types, said Wallace: blue-water rivers, black-water rivers, and white-water rivers, by which Wallace actually meant the distinctive pale yellow-olive, the color of the Amazon's main stream. Sediment is the solid material, both mineral and organic, that is suspended in water and is being transported from its site of origin. The Amazon normally takes its colors from the sediments of the lands the waters flow through. A rocky-sandy country, said Wallace, produces clear waters. Clayey soil gives up the yellow-olive sediment. The Rio Branco (meaning white in Portuguese) Wallace described as "a milky color mixed with olive," as if carrying chalk in solution. On the Rio Branco's banks Wallace saw "considerable beds of the pure white clay which occurs in many parts of the Amazon." Another fat tributary, the Toncantins, runs mostly over volcanic rock; its waters are "beautifully transparent," said Wallace.

The most curious color phenomenon, however, was the great river's black waters, like the celebrated Rio Negro (meaning black in Portuguese). The causes of this phenomenon "are not very obscure," Wallace explained, but came from above ground level. "It appears to me to be produced by the solution of decaying leaves, roots, and other vegetable matter." The jungle river thus adopts the color not of the sediment but of the accumulated detritus that it sweeps down toward the sea. Wallace thought it might have something to do with the Rio Negro becoming stagnant at certain times of the year because of differential rain patterns on the Negro and lower Amazon.

The colors of the Amazon brought Wallace to investigate the sediment and substrata. He found the "almost perfect flatness" of the Amazon Valley its single most striking geological fact. No mountains or even slightly elevated plateaus rise from the plain until you reach the abrupt peaks of the Andes. Wallace's impression was that "here we see the last stage of a process that has been going on, during the whole period of the elevation of the Andes"—the gradual filling in of what was once the granite bottom of the sea with sediment brought down by rivers from the Andes Mountains. Wallace complained that the Amazon's geology was difficult to study: the flat terrain and heavy forest cover made natural sections "comparatively scarce." The king of collectors did not find a single fossil in the Amazon. "Not even a shell or a fragment of fossil wood, or anything that could lead to the conjecture as to the state in which the valley existed at any former period," he wrote in chagrin.

The general geological story, however, was clear enough to him. The base was granite. "Granite seems to be, in South America, more extensively developed than in any other part of the world." Wallace met with granite over the whole upper part of the Amazon and Rio Negro, and far up the Uapes River toward the Andes. On the Rio Negro he described the granitic formations as "spread out in immense undulating areas," the hollows of which were gradually filled with sediment deposits to form smooth alluvial tracts of clays within or on top of the granite base. On these beds of clay grew the forests. The "marvelous regularity of the surface" Wallace attributed not only to the creation of sedimentary beds by erosion but to a "shaking and as it were, levelling" of the sediment by volcanic activity.

Having reviewed the Amazon's geography and geology, we are now ready to meet some of the Amazonian wildlife in Professor Wallace's collection—the real Wallace, one is tempted to say. He actually found the Amazon deficient in large animals and agreed with Darwin's Patagonian conclusion that large mammals are generally better adapted to grazing open plains than thick rain forest. Wallace described three small species of deer as the sole representatives of the ungulates; the tapir, which is the only extant New World pachyderm; two species of wild hogs; two or three species of large cats; and "those singular creatures, the sloths, armadillos and ant-eaters." Wallace met with twenty-one species of anteaters.

The monkeys, said Wallace, were "the only animals found in any numbers" in the Amazon Valley, "the only mammalia that give some degree of life to these trackless forests, which seem peculiarly fitted for their development and increase." Note the use of the word "fitted," as in the phrase that later defined evolution for the public, "survival of the fittest." Only here the trees are fitted to the monkeys, not the other way around. Wallace's frequent use of the word "peculiar" also requires a brief explanation. Although today peculiar means "strange," "odd," or even "weird," all with negative connotations, the word used to be primarily employed (and still often is used in Britain) to mean "singular" or "unique." Wallace is examining here the unique relationships between a habitat and the species of animals that evolve and survive in it; in this case, tree canopy covering a thousand miles has spawned a highly diverse, numerous, and successful fauna of mammals that live in trees.

Of the monkeys he was able to examine, Wallace said the howlers, forming the genus *Mycetes*, "were the largest and most powerful." Certainly anyone who has been to contemporary Neotropical jungles will also grant the howling monkeys first place as most prominent. The stereotypical jungle image of the gorilla beating its chest or the humanoid chimpanzee give way in the New World to yearning twilights, pounding with the echoing cries of the howlers. Wishing to find out how they make their booming, rolling howls, "which appears as if a great number of animals were crying in concert," Wallace apparently dissected one and found "a boney vessel situated beneath the chin and a strong muscular apparatus in the throat, which assists in producing the loud rolling noise from which they derive their name." The extra bone and muscle must support the unusually strenuous movement of the voice box necessary to emit and sustain such a big noise. Howls last for minutes on end.

Another infamous mammal of the Amazon that Wallace came to know was the vampire bat, *Phyllostoma hastatum*, whose diet of blood became the literary vehicle of Bram Stoker's Count Dracula, then ten thousand horror movies and books. Wallace cited this common bat as doing "much injury to the horses and cattle by sucking their blood; it also attacks men, when it has opportunity." Unlike Charles Waterton, who used to sleep with his big toe sticking out of the mosquito netting of his hammock in the hope that a bat would bite him, but who was never bitten, Wallace was attacked twice, once on the tip of the toe, the other time on the tip of his nose. "In neither case did I feel anything, but awoke after the operation was completed: in what way they effect it is still quite unknown."

Now it is quite well known: the vampire bat makes a clean and painless surgical slash with its teeth, then laps up the blood with its tongue. The bat's saliva contains anticoagulants to prevent the blood from clotting while the bat is feeding. That is why, as Wallace noted, the bite of the vampire bat is so slow to heal. What was also unknown to Wallace was how selective his facts about the bats were in describing only vampires. The Amazon rain forest is a paradise for fruit-eating bats. Their importance in the pollination and reproduction of jungle plants is so great that they are now recognized as crucial to the rain forest's survival. But probably because they are nocturnal and airborne, Wallace fails to mention them. Only recently have biologists begun to understand the wonderfully complex nexus of relations between the fruit-, pollen-, and nectar-eating bats and the many plants that vie for their attention in the tropical night.

Although camouflage and cryptic coloring in general were the specialty of his colleague Bates, Wallace had a few words to say about the spotted jaguar, or *onca*, as it is called in Brazil, the Amazon's king of cats. The jaguar frequently conceals itself on a tree branch, where it blends in with the sunstruck greenery, waits for prey to pass underneath, then simply drops down. As if this big cat did not have enough advantages in the jungle struggle for survival in its invisibility, strength, speed, fierceness, and cunning, Wallace repeated the "general belief among the Indians and the white inhabitants of Brazil that the *onca* has the power of fascination."

What on earth did he mean? "Fascination," in one of its older meanings, is the power to deprive of resistance by terror or even magic. A person had told Wallace of seeing a jaguar standing at the foot of a high tree, staring down a monkey. The monkey crept slowly down the tree, crying piteously the whole time, and finally dropped itself at the very feet of the jaguar, "which seized and devoured it." Whether exaggerated or imagined, such stories, Wallace reported, were commonplace in the Amazon Valley, and universally believed. Wallace doesn't say he believed them, but he doesn't say he disbelieved them, either.

As for birds, Wallace declared them so numerous and diverse that he could mention only a few of the more interesting and beautiful ones. Among tanagers, trogons, and tyrant shrikes, cotingas and oropendolas (which are orioles that build striking hanging-basket nests), Wallace thought the parrots and toucans "perhaps the most characteristic" of the Amazon jungle, abounding in species and more frequently encountered than any other birds. At least they were the most easily recognizable, the toucans being instantly identified by their large horny bills, often colored in brilliant hues. Wallace praised the "most curious and beautiful" of the sixteen toucan species he collected, the curl-crested aracari, which has a glossy crest of horny black curls. What purpose these horny bills serve the toucans remains something of a puzzle. Ornithologists speak of them as aiding identification in selecting a mate, but when it comes to such bursts of art, it seems to me we must look beyond mundane utility to nature's caprice.

Wallace took note of the hummingbirds, which although almost entirely confined to tropical America, he found in no great numbers in the lowlands. Still, he could end his short section on birds: "Probably no country in the world contains a greater variety of birds than the Amazon Valley. Though I did not collect them very assiduously, I obtained upwards of five hundred species, a greater number than can be found all over Europe. . . . Anyone collecting industriously for five or six years might obtain near a thousand different kinds."

The animals and plants of the Amazon jungle are peculiar and fascinating. But what about Wallace's idea to make an Amazonian collection that would accomplish no less a scientific feat than to shed light on the mechanism of evolution?

"There is no part of natural history more interesting or instructive than the study of the geographical distribution of animals," Wallace begins his discussion with circumspection. Wallace thought "there must be some kind of boundary which determines the range of each species; some external peculiarity to mark the line which each one does not pass." What external factors, Wallace asked, might limit the range of animals? Climate? Seas? Great mountain chains? Deserts? "It is well known," he continued, "that countries possessing a climate and soil very similar may differ almost entirely in their productions [i.e., fauna]." Look, for example, at Europe and North America—similar regions, different fauna. South America, Wallace said, contrasts sharply with the African shoreline. Australia, too, shares latitudes and climates with parts of South America, yet Australia has its own endemic creatures.

The physical environment can indeed block the spread of land species. Seas can. So can mountain chains. Wallace maintained that the Amazon was an example of a large river limiting range, though mostly in the case of animals that could not fly or swim. But there must be many other kinds of natural boundaries, and Wallace noted that each continent had "well-marked smaller districts" where the distribution of species appeared to depend not on blocking but on climate. In addition, Wallace recognized still smaller units, which today would be called local or micro ecosystems. Almost every such area, he recognized, "has peculiar animals found nowhere else." He was approaching the idea of niches: those narrow places in nature where species can succeed by specializing.

Wallace gave a fine example in describing the habitat that results from the seasonal differential of the Amazon River. Like all tropical rivers, the Amazon is subject to an annual rise and fall of great regularity, which Wallace tabulated as about fifty feet at Barra. The Amazonian waters begin to rise with the rains in December–January, and reach their crest in June. Wallace calculated that seventy-two inches of rain—six feet!—fall annually in the Amazon Valley.

At flood season the river constantly overflows its low, flat banks. These floodplains can spread forty to fifty miles into the jungle, fostering a unique flooded forest then called *gapo*, teeming with life forms, where thousands of varieties of nut- and seed-eating fish essentially "farm" the tropical trees by dispersing their seeds for them. These vast flooded tracts, submerged for six months of the year, are one of the most distinctive features of the Amazon Valley. They form a secluded water world the natives use to canoe and fish for great distances, away from the currents of the main river.

Surely the most interesting group in regard to niches would be the most numerous, i.e., the insects—"the countless tribes of insects that swarm in the dense forests of the Amazon," as Wallace wrote. At Pará he had collected "six hundred distinct kinds" of butterflies within a day's walk of the city. Among the Coleoptera and Hymenoptera (beetles), the Lepidoptera and Heliconiida (butterflies), Wallace was supposed to crack open the case of evolution, revealing stunning facts concluded from his Amazon Valley collection. In his review, one can plainly hear the voice of the collectors' demon breaking through: "At Santerem I had increased my collection to seven hundred species, at Barra to eight hundred, and I should have brought home with me nine hundred species. . . ."

But it is just here that Alfred Russel Wallace has nothing more to say—and here we come to the surprising and sad conclusion of his exploration of the Amazon Valley.

At the end of four years collecting in the Amazons, Wallace called it quits. He was weakened by malaria—his brother, who had come out to help him collect, died of it—and ready to return to England to see what he had. He went back to Pará at the mouth of the river, gathering and packing up all his collections. It was a massive undertaking. Box after box of specimens, not to mention his collection of one hundred live animals—reduced to thirty-four crates before embarking for England, mainly due to accidents. Wallace's prize toucan was drowned, for example, when it fell out of a canoe on the journey downriver; toucans are notoriously bad flyers.

Wallace booked passage to London on the brig *Helen*, a 235-ton ship carrying a cargo of rubber and cocoa. When his entire treasure was loaded, Wallace left the Amazon on July 12, 1852. Two days after departure the fever returned, and Wallace, suffering also from seasickness, spent the next three weeks mostly in bed in his cabin.

On the morning of August 6, while Wallace was below reading, the captain came down and announced, "I'm afraid the ship's on fire." Wallace raced up to the deck in time to see "a dense vapoury smoke issuing from the forecastle." For hours the crew and passengers fought the blaze. All attempts to put it out failed. The flames spread. Wallace watched in the detachment of shock as his live animals were burnt or suffocated. Within hours, fire consumed the *Helen* and sent her to the bottom. Wallace watched from a long boat as his Amazon Valley collection went down with her. He was rescued after nine days adrift at sea, about two hundred miles from the island of Bermuda.

Only after he recovered from the trauma did Wallace realize the greatness of his loss: "With what pleasure had I looked upon every rare and curious insect I had added to my collection! How many times, when almost overcome by the ague, had I crawled into the forest and been rewarded by some unknown and beautiful species! How many places, which no European foot but my own had trodden, would have been recalled to my memory by the rare birds and insects they had furnished to my collection!"

But his collection was gone. "I had not one specimen to illustrate the unknown lands I had trod, or to call back the recollection of the wild scenes I had beheld."

So Alfred Russel Wallace did not solve the riddle of evolution in the Amazon Valley. The idea of natural selection came to him years later in a brainstorm while he lay wasted again with malaria in Malaysia. In the end, his years of rational collection of facts prepared Wallace for his febrile moment of illumination. As Pasteur observed, chance favors the prepared mind.

5

Henry Walter Bates on the Rivers Amazon

IN HENRY WALTER BATES we meet again, as with his partner Wallace, the nineteenth-century tropical naturalist as collector. In his expedition to the Amazon River Basin lasting eleven years—ended only because he was suffering from malaria—Bates gathered such impressive specimens as to make him the all-time champion of collectors in the American tropics. And he got his collection back to England intact. In all, he had amassed nearly fifteen thousand creatures, most of them insects. Some eight thousand of these were new to science at the time. Not without self-effacing pride in these numbers, Bates tells us in his memoir,

A Naturalist on the Rivers Amazon, that his four-and-a-half-year stay on the upper Amazon River yielded such a large number of new species (more than three thousand) from the one locality of Ega that "the name of my favourite village has become quite a household word amongst a numerous class of Naturalists."

Bates made 550 species of butterflies at Ega—where, he informs us, 18 species of swallowtails were found within ten minutes' walk of his house, surely a new record, though in an obscure category. The neighborhood of Pará, Bates goes on, yielded 700 species of butterflies, compared with a paltry 66 for the entire British Isles. He obtained 360 birds, 140 reptiles, 120 fish, 35 mollusks, and an incredible 14,000 species of insects. The forest creatures must have quaked when they saw Bates coming with his gun and net. Yet no one fell headlong in rapture with the tropical forests as much as the man whose project there was to bring back dead specimens of its inhabitants.

It is usually said of Bates and the other naturalist collectors that they were merely acting in the line of scientific duty, filling in the many blanks in taxonomic classification. But the true collector, as I have indicated in describing Wallace, is also operating under the impulse of obsession. The urge to amass an ever greater number, or a complete set, of some objects can take powerful hold of the human soul. Like the greedy man with his wealth, the collector seeks to agglomerate more of something than anyone else. The collector, however, also takes pride in learning about the material object of desire. Through his entomological collecting activities, Bates became the world's authority on insect fauna of the Amazons.

Bates was born at Leicester, England, on February 8, 1825, son of a small textile manufacturer whose own father had been a millhand. Unlike Waterton, the eccentric aristocrat, Bates was born into the lower middle class, where the future held no university education, only a drone's life working in a factory or office.

His formal education ended, like Wallace's, at age thirteen, when he was apprenticed as a clerk in the manufactory, working thirteen hours a day. But by this time he had already caught the collecting bug and spent all the time he could gathering insects, especially beetles and butterflies. How fascinating the intricate workings of the beetle! What marvels the colorful wings of the butterflies! The young naturalist with his collection spread before him had already escaped the dull world of grey smokestacks and account books. Bates was, again like Wallace, an autodidact, who also taught himself Greek and Latin and to play the guitar.

One day in the Leicester public library Bates had a chance encounter with another young working bloke, who was also fascinated by natural history. This was, of course, Alfred Russel Wallace. The two young naturalists were outwardly much alike. Wallace later recalled in his autobiography the first meeting of these two enthusiastic entomologists, which was to change their lives: "He asked me to see his collection, and I was . . . surprised when I found that almost all had been collected around Leicester, and that there were still many more to be discovered. If I had been asked before how many kinds of beetles were to be found in any small district near a town, I should probably have guessed about fifty or at the outside a hundred, and thought that a very liberal allowance. But I now heard that there were probably a thousand different kinds within ten miles of the town."

The two lads soon became friends and collecting companions. Together they rambled through the countryside on Sundays and holidays, sharing their enthusiasms like naughty boys engaged in mischief. They discussed such subversive subjects as Lyell's *Principles of Geology* and Darwin's recently published account of the voyage of the *Beagle*. Soon

Henry Walter Bates collected almost fifteen thousand creatures, about half of them new to science, during his years on the Amazon.

enough they were fantasizing about escaping England for the tropics, to collect in virgin terrain. Visions of some vast uncollected land, where the sun was hot and collecting never ended, danced in their heads. They yearned to get away from the stifling cities, away from their boring jobs, away from the stultifying conformity of England, with its class-ridden limitations to young men and women of enterprise.

With the publication in 1847 of the American lepidopterist W. H. Edwards's account of his expedition to the Amazon River, the dam burst. Wallace and Bates decided to throw it all over and head out for Brazil together. They would pay their way by collecting specimens for sale to museums and university collections—for the great public institutions had also fallen to the collecting demon. They would become, as it were, professional collectors. Wallace was twenty-five, Bates twenty-three.

Confronted by a territory so almighty vast, foreign, and largely unexplored as Amazonia in 1848, any naturalist had first to make some basic decisions about what to accomplish and how to go about it. After an initial period getting used to the climate, learning Portuguese, and collecting in the forest near Pará, Wallace and Bates went their own ways. The different journeys they pursued as naturalists were undoubtedly to some extent a reflection of their individual temperaments. Wallace set out to explore the upper reaches of the basin along the Rio Negro above Manáos, where no whites had ventured before. Bates never went far from the main stream of the Amazon. He preferred to settle in one place where the collecting was good, and to remain there until its possibilities were exhausted. "I wished to explore districts at my ease," he tells us. In this way, Bates passed the greater part of eleven years in just three regions. In each case he made a town on the Amazon his headquarters, and from that point explored the surrounding region. He was nearly three years at Pará near the mouth of the river; three and a half years at Santarem, a few hundred miles upstream at the mouth of the Tapajós; and the last four years at Ega, which is today called Tefé, at the mouth of the Rio Tefé. This methodology well suited his temperament. He had found places where more than half the creatures were unknown to science. Each day he added a new species of butterfly or beetle, wasp or ant, to his expanding collection. Why move when the collectors' demon could be fed so richly?

Bates has left us a fascinating sketch of how the naturalist works from day to day in the tropical forest, adjusting his schedule according to the diurnal rhythms governing human as well as animal species: "We used to rise soon after dawn, when Isidoro [his assistant] would go down to the city, after supplying us with a cup of coffee, to purchase fresh provisions for the day. The two hours before breakfast were devoted to ornithology. At that early period of the day the sky was invariably cloudless (the thermometer marking 72 or 73 degrees Fahr.): the heavy dew of the previous night's rain, which lay on the moist foliage, becoming quickly dissipated by the glowing sun. . . . The birds were all active; from the wild-fruit trees, not far off, we often heard the shrill yelp of the Toucans. Small flocks of parrots flew over on most mornings. . . . After breakfast we devoted the hours from 10 a.m. to 2 or 3 p.m. to entomology; the best time for insects in the forest being a little before the greatest heat of the day.

"The heat increased rapidly towards two o'clock (92 and 93 degrees Fahr.) by which time every voice of bird or mammal was hushed; only in the trees was heard at intervals the harsh whirr of a cicada. The leaves, which were so moist and fresh in early morning, now become lax and drooping; the flowers shed their petals. Our neighbours, the Indian and Mulatto inhabitants of the open palm-thatched huts, as we returned home fatigued with our ramble, were either asleep in their hammocks, or seated on mats in the shade, too languid even to talk. On most days in June and July a heavy shower would fall sometime in the afternoon, producing a most welcome coolness. . . .

"Towards evening life revives again, and the ringing uproar is resumed from bush and tree. The following morning the sun again rises in a cloudless sky, and so the cycle is completed; spring, summer, and autumn, as it were, in one tropical day."

Bates was meticulous, a detail man. Like many entomologists, he preferred to look at the tiny mechanisms of life, the microworld of the jungle insect. Here he is unsurpassed, exploring the subterranean galleries of the *sauba*, or leafcutter ants, "their processions . . . like a multitude of animated leaves on the march," as they clipped and carried away immense quantities of green leaves:

"In some places I found an accumulation of such leaves, all circular pieces, about the size of a sixpence, lying on the pathway, unattended by ants, and at some distance from any colony. Such heaps are always found to be removed when the place is revisited the next day. In course of time I had plenty of opportunities of seeing them at work. They mount the tree in multitudes, the individuals being all worker-minors. Each one places itself on the surface of a leaf, and cuts with its sharp scissor-like jaws a nearly semicircular incision on the upper side; it then takes the edge between its jaws, and by a sharp jerk detaches the piece. Sometimes they let the leaf drop to the ground, where a little heap accumulates, until carried off by another relay of workers; but, generally, each marches off with the piece it has operated upon, and as all take the same road to their colony, the path they follow becomes in a short time smooth and bare, looking like the impression of a cartwheel through the herbage."

With a clear eye, an active mind, and a felicitous pen, Bates describes the itchy activities of the *garapatos*, or ticks, the torments of the mosquitoes, the terror of the army ants, the insufferability of the *bête rouge*, or chiggers, as well as the wonder of the carnivorous beetles, furnished, as he observed, with "a beautiful contrivance" for enabling them to cling to and run over smooth or flexible surfaces, such as leaves: "Their tarsi or feet are broad, and furnished beneath with a brush of short stiff hairs; whilst their claws are toothed in the form of a comb, adapting them for clinging to the smooth edges of leaves, the joint of the foot which precedes the claw being cleft so as to allow free play to the claw in grasping."

One of Bates's constant curiosities is the way tropical insects employ mimicry to defend themselves against predation. For when the total biomass of tropical fauna is weighed, the proportion held by the insects wins hands down, making them the chief food source for carnivores. Bates describes, for instance, the hummingbird moth, several of which he shot by mistake while thinking to collect hummingbirds:

"This moth (*Macroglossa titan*) is somewhat smaller than humming-birds generally are; but its manner of flight, and the way it poises itself before a flower whilst probing it with its proboscis, are precisely like the same actions of humming-birds. . . . The resemblance between this hawk-moth and a humming-bird is certainly very curious, and strikes one even when both are examined in the hand. Holding them sideways, the shape of the head and position of the eyes in the moth are seen to be nearly the same as in the bird, the extended proboscis representing the long beak. At the tip of the moth's body there is a brush of long hair-scales resembling feathers, which, being expanded, looks very much like a bird's tail. But, of course, all these points of resemblance are merely superficial. . . . The analogy between the two creatures has been brought about, probably, by the similarity of their habits, there being no indication of the one having been adapted in outward appearance with reference to the other."

Bates was probably wrong on this point. Hummingbird moths, like hawkmoths, which have eyespots on their wings resembling the eyes of hawks, mimic birds in order to startle and deter potential predators. Bates himself was later to describe for the first time a closely allied variety of mimicry in tropical butterflies. Although they look merely marvelous to us humans, some butterflies are beautifully colored in order to signal predators that they don't taste good, or are poisonous, and thus not to be considered a food item. In Amazonia the combination of black, yellow, and orange coloration on the *Heliconid* butterfly warns predators to stay away. Bates found that another butterfly, the *Dismorphia astyocha*, which has evolved to closely resemble the female *Heliconius* with black, yellow, and orange coloration, is actually harmless to eat. Many good-tasting butterflies gain protection from predators by resembling the bad-tasting ones—a kind of sheep in wolf's clothing syndrome, which has come to be called Batesian mimicry. It is one of the tropical jungle's secret languages.

But Bates by no means limited his writings to the insects. If *The Naturalist on the Rivers Amazon* has endured as a classic of tropical natural history, it is principally for the romantic miniature landscapes and scenes Bates paints of traveling by canoe through this extensive and amazing tropical wilderness. Bates journeyed at the end of an era, just before the world was divided into professional scientific specialties. He felt free to write about geography, climate, the natives, the villages and plantations, and, of course, the rivers and the jungle. Here he is, for example, floating on the Rio Tocantins, awaiting the flood tide to cross the channel to Cameta:

"I fell asleep about ten o'clock, but at four in the morning John Mendez woke me to enjoy the sight of the little schooner tearing through the waves before a spanking breeze. The night was transparently clear and almost cold, the moon appeared sharply defined against the dark blue sky, and a ridge of foam marked where the prow of the vessel was cleaving its way through the water. The men had made a fire in the galley to make tea of an acid herb, called *erva cidreira*, a quantity of which they had gathered at the last landing-place, and the flames sparkled cheerily upwards. It is at such times as these that Amazon travelling is enjoyable, and one no longer wonders at the love which many, both natives and strangers, have for this wandering life."

Or here, describing the weird uproar of life that sometimes takes place at sunset in the tropical forest: "The noises of animals began just as the sun sank behind the trees after a sweltering afternoon, leaving the sky above of the intensest shade of blue. Two flocks of howling monkeys, one close to our canoe, the other about a furlong distant, filled the echoing forest with their dismal roaring. Troops of parrots, including the hyacinthine macaw we were in search of, began then to pass over; the different styles of cawing and screaming of the various species making a terrible discord. Added to these noises were the songs of strange Cicadas, one large kind perched high on the trees around our little haven setting up a most piercing chirp. . . . The uproar of beasts, birds, and insects lasted but a short time: the sky quickly lost its intense hue, and the night set in. Then began the tree-frogs—quack-quack, drum-drum, hoo-hoo; these accompanied by a melancholy night-jar, kept up their monotonous cries until very late."

These views and experiences leaven the sometimes tedious accumulation of specimens and facts that was Bates's work. Yet for the most part he toiled with his head down, piling fact upon fact, observation upon observation, species upon species. Only rarely does he break into the abstract generalization, the free speculation, the spontaneous idea or philosophical conjecture. In a book of almost four hundred pages recounting the travels and studies of eleven years in Amazonia, we can almost count on one hand the times Bates takes up such broader subjects as evolution, adaptation, or the modification of species. When he does, one has the sense that he has looked up half-grudgingly from his butterfly net or beetle chloroform, and tipped his hat to Charles Darwin, getting back to his collecting as soon as possible.

Bates would probably not have written *The Naturalist on the Rivers Amazon* had it not been for the friendly insistence of Darwin. On his return to England in 1859, Bates was slow to recover from his malarial weakness, exacerbated by malnutrition. He gradually gave up the idea of making a book from the voluminous notes of his Amazon adventures. Two years elapsed before he made Darwin's acquaintance. Darwin was enthusiastic in his praise of Bates's mission to the Amazon, and strongly urged him to write a book of his travels. Could anyone refuse Charles Darwin?

Bates began immediately, and Darwin continued his encouragement to the younger naturalist, praising the book after publication as second only to Humboldt's in its descriptions of tropical forests. Could the tutelage of the great theorist have had something to do with Bates's almost bashful mentions of uniformitarian principles, evolution, and adaptation? Perhaps so.

Even less does Bates speak of himself directly. His collector's vocation seems to have gone along with a mild, circumspect, solitary personality. He shunned not only the public spotlight of scientific debate but exposure of his own life in print. He must have had a great tolerance for loneliness. In one Amazonian village, he writes, "I led a solitary but not unpleasant life; for there was a great charm in the loneliness of the place. The swell of the river beating on the sloping beach caused an unceasing murmur, which lulled me to sleep at night, and seemed an appropriate music in those midday hours when all nature was pausing breathless under the rays of a vertical sun."

In one unusually revelatory passage he refers to himself as a "solitary stranger" on "a strange mission"—an apposite self-evaluation. And in another he quietly admits having had no contact with anyone outside the forest for a year, and having read and reread every copy of his *Atheneum* magazine (including saving the advertisements for special occasions), that nature alone was not enough to make life worthwhile.

But for the most part, though Bates takes us to the headwaters of the Amazon, we never make much headway on the question of what possesses a man to break all ties with home, country, and family, and bury himself for years in the jungle of a foreign continent. Only Bates's reaction upon leaving Brazil gives us an idea of his emotional attachment: "On the evening of the third of June, I took a last view of the glorious forest for which I had so much love, and to explore which I had devoted so many years." The saddest hours of his life, Bates tells us, were those spent on board ship, when he realized the last links connecting him to his "naturalist's paradise" were broken. Before him suddenly loomed images of the English climate, scenery, and way of life. "Pictures of startling clearness rose up of the gloomy winters, the long grey twilights, murky atmosphere, elongated shadows, chilly springs, and sloppy summers; of factory chimneys and crowds of grimey operators, rung to work in the early morning by factory bells; of union workhouses, confined rooms, artificial cares and slavish conventionalities."

Too late, Bates realized he was returning to the freeman's hell, the tropical naturalist's prison. The society that had made him an outdoorsman and naturalist would swallow him back up. The conventionality and stuffiness of middle-class life, the class-consciousness and meanness of British society, the grey bleakness of industrialism—these were the things that turned Englishmen into naturalist collectors and sent that small swarm of Brits under the equator—men, and women too, who could not wait to fill their lungs with the open air of tropical rivers, feel the equatorial sun burn their skin, fill their eyes and minds with the great carnival of tropical life, and explore a continent where freedom of action and democracy of science were to be had in all directions, for thousands of verdant miles.

6

Thomas Belt: A Naturalist in Nicaragua

DANIEL JANZEN, the recognized dean of modern tropical biologists, wrote in his 1985 foreword to a new edition of *The Naturalist in Nicaragua*: "Thomas Belt lived his forty-five years matter-of-factly close to nature. It is difficult for us to appreciate how close, for today we travel in an afternoon the distance it took Belt nineteen days to cover on muleback. When you ride a mule for nineteen days, you have time for the contemplation of your surroundings, and your surroundings have the time to contemplate you. This is not to suggest that perching on a slogging mule in the blazing sun or driving rain, with no food in your stomach, will lead automatically or even at all to a careful, enthusiastic, interpretive examination of nature.

"But Belt was an engineer, and he went at nature with the attention to detail one would expect of a man who did not want his bridges or mine shafts to collapse. A man whose life depends on getting a timber baulk at just the right point and angle in a tunnel is someone preconditioned to notice quite carefully how a bee enters a flower."

Thomas Belt was widely recognized during his short life as something special in natural history. His engineering background led him to brilliant descriptions of many designs and systems in tropical nature, evolution, and earth science in general. As someone whose profession was figuring out where and how best to mine the earth, Belt was exceptionally strong in geology, physical geography, and entomology—insects being the demiengineers of the natural world. Belt was a true Victorian, a British purveyor of progress to the South, burdened by the civilization he bore but bearing it, nonetheless, with a stiff upper lip. His book is a collection of scientific essays on subjects Belt studied and contemplated for many years at close range in the line of duty as a gold mining engineer for a British company at Chontales, on Nicaragua's Atlantic coast—animal mimicry, the interactions of bullthorn acacias with ants, the evolution of special adaptations for defense, and the glaciation of Central America during the Ice Age. Belt amalgamated his scientific material with a thready travel narrative of his muleback sojourns through Nicaragua.

Unlike Bates, with whom Belt was friendly in later days when Bates became editor of the *London Geographical Society Bulletin*, Thomas Belt did not shy away from wider conclusions and theoretical speculations. His book on Nicaragua was published in 1874, fifteen years after Darwin's *Origin of Species*. Belt not only enthusiastically em-

Thomas Belt encounters a jaguar in the jungles of Nicaragua.

braced Darwin's theory of natural selection but placed his own conclusions at the service of what became known as Darwinism. In the event, the protégé outdid the mentor, and at times Belt's Darwinian dedications sound suspiciously akin to the "iron laws" of human science that Lenin proclaimed after Marx was long dead and buried in England. Ever since Darwin, the work of all tropical biologists has essentially been to prove the Big Man's theory by filling in the many nontheoretical blanks. Belt was only the first in a long line of increasingly professional, academically trained researchers.

A chip falling so close to the old block, Belt can be slavish to his master. In the facsimile edition of *The Naturalist in Nicaragua*, for example, is an illustration of the heads of motmot birds that is almost identical to Darwin's Galápagos Islands finches in *Voyage of the Beagle*. Homage verging on plagiarism doesn't end with the artwork. Belt says of the motmots, "They all have several characters in common, linked together in a series of gradations. One of these features is a spot of black feathers on the breast." He goes on to describe the small variations among species of the motmot family. These are the gradations of the bills of Darwin's finches. Belt is so eager to rush ahead to bolster Darwin's "supposition of descent of different species from a common progenitor" that he fails to tell his readers anything about the really interesting feature of motmots, which is that they have long tails with a featherless section above the tip, at which point there is a circle of feathers, making the motmot's tail look like a tennis racket, which the bird swings back and forth when on perch, perhaps practicing its forehand.

The Nicaragua Thomas Belt traveled through from 1868 to 1872 can be divided into three longitudinal zones running northwest to southeast. The easternmost, or Atlantic, zone is an extensive lowland coastal plain known as the Mosquito Coast, a corruption of Miskito, the name given to the Miskito Indians. This is not an ancient aboriginal tribe but seems to have formed as a culture in the seventeenth century, when some forest dwellers moved out to the coast to trade with the English buccaneers who hid out along this coast in order to waylay Spanish galleons carrying the New World's gold to Europe. Intermarriage accounts for the blue-eyed henna-haired Indians one meets in the Mosquitia, with names like Franklin Thompson and Brooklyn Sawyer.

A lively trade sprang up then, with the natives supplying turtle meat and other provisions to the pirates in return for guns known as muskets. The Indians employed these weapons to conquer or eliminate the other natives dwelling around the many rivers of the interior of the Atlantic zone, and a new dominant trading culture formed around turtle hunting and firearms. The use of British muskets distinguished these Indians, and it is presumed that the word was corrupted from musket to Miskito, and had nothing whatsoever to do with biting insects.

The Mosquito Coast and the lowlands behind, Belt tells us, were covered, then as today, by a great unbroken rainforest: "The Atlantic forest, bathed in rains distilled from the northeast trades, is ever verdant. Perennial moisture reigns in the soil, perennial summer in the air, and vegetation luxuriates in ceaseless activity and verdure, all the year around. . . . Unknown the cold sleep of winter; unknown the lovely awakening of vegetation at the first gentle touch of spring."

It was in the Atlantic rain forest zone of Nicaragua that Belt made some of his most original observations. Of hummingbirds, for example, he shows that these iridescent jewels of the bird world are not nectar sippers or honeysuckers, as most people believe. Their tongues, the engineer informs us, are made of a semihorny material for the lower half of their length, and are cleft in two. The two halves are laid flat against each other when at rest, but they can be separated at will—a design strikingly like "a delicate pliable pair of forceps, most admirably adapted for picking out minute insects from amongst the stamens of flowers."

This kind of subject is Belt's forte. When he moves from the physical mechanism to the behavioral, he often has problems. His puzzled account of the "squeeking" hummingbirds, for example: "There were four or five other small ones [hummingbirds] that we used to call squeekers, as it is their habit for a greater part of the day to sit motionless on branches and every now and then to chirp out one or two shrill notes. At first I thought these sounds proceeded from insects, as they resemble those of crickets; but they are not so continuous. After a while I got to know them, and could distinguish the notes of different species. It was not until then that I found out how full the woods are of humming-birds, for they are most difficult to see when perched amongst the branches, and when flying they frequent the tops of trees in flower, where they are indistinguishable. . . . My conclusion, after I got to know their voices in the woods, was that the humming-birds around Santo Domingo equalled in number all the rest of the birds together, if they did not greatly exceed them. Yet one may sometimes ride for hours without seeing one."

Belt's conclusion is completely wrong, but it would be eighty years before Alexander Skutch solved the mystery of the "squeeking" hummingbirds. The reason they seemed to Thomas Belt to be as numerous as all other birds put together, yet sometimes not seen for hours at a stretch in the forest, is that male hummingbirds congregate in what is called a lek. The males, which cannot sing worth a damn, gather in one small area of the forest, often within sight of one another and certainly within earshot, and engage in these squeeking competitions for hours on end. Skutch felt certain that hummingbird leks are pre-mating displays; essentially the males get together to compete vocally for the females. Leks tend to repeat year after year, generation after generation, in the same place. It is not known whether the females find the squeeky male vocalizations attractive or judge by loudness or some other measure. Skutch also gathered evidence that male hummingbirds sometimes gather in leks out of mating season, and he suggested that they may be competing for the sheer fun of it, engaging in a social activity like sports.

Of the relation between special defenses and mimicry in the tropical moist forest, Thomas Belt has much to say, nearly all right on the mark, including the admirable reminder to his readers that when we say one insect "mimics" or "imitates" another, "it is as if it were a conscious act." What advocates of natural selection meant by mimicry, Belt said, were "deceptive resemblances . . . brought about by varieties of one species somewhat resembling another, having special means of protection (bad taste, toxicity, a stinger, camouflage for concealment), and preserved from their enemies in consequence of that unconscious protection. The resemblance, which was perhaps at first only remote, is supposed to have been increased in the course of ages by the varieties being protected that more and more closely approached the species imitated in form, color, and movements."

Just as Charles Waterton reached some of his best insights swinging in his hammock, Belt made many of his best discoveries bouncing in the saddle. It was his habit, he tells us, to ride along on his burro or *caballo* carrying a net fixed to a short stick, catching insects off the leaves as he passed along, without stopping. After one such capture-on-the-hoof Belt observed in his net what appeared to be a small, black, stinging ant. He killed it to avoid the sting, but, looking closer, he found it wasn't an ant after all but a small spider that perfectly imitated an ant, even to the point of waving about its two forelegs like antennae. Belt wrote:

"Ant-like spiders have been noticed throughout tropical America and also in Africa. The use that the deceptive resemblance is to them has been explained to be the facility it affords them for approaching ants on which they prey. I am convinced that this explanation is incorrect so far as the Central American species are concerned. Ants, and especially the stinging species, are, so far as my experience goes, not preyed upon by any other insects. No disguise need be adopted to approach them, as they are so bold that they are more likely to attack a spider than a spider them. Neither have they wings to escape by flying, and generally go in large bodies easily found and approached. The real use is, I doubt not, the protection the disguise affords against small insectivorous birds. . . . Stinging ants, like bees and wasps, are closely resembled by a host of other insects; indeed, whenever I found any insect provided with special means of defense I looked for imitative forms, and was never disappointed in finding them."

The phosphorescent species of fireflies, equally distasteful to predators, Belt found also much mimicked by other insects, especially by certain cockroaches. He noted that instead of hiding in crevices and under logs like their brethren, the mimetic roaches rested during the day exposed on the surfaces of leaves, in the same manner as the fireflies they mimicked:

"All the insects that have special means of protection, by which they are guarded from the attacks of insectivorous mammals and birds, have peculiar forms, or strongly contrasted, conspicuous colours, and often make odd movements that attract attention to them. There is no attempt at concealment, but, on the contrary, they appear to endeavour to make their presence known. The long narrow wings of the Heliconii butterflies, banded with black, yellow, and red, distinguish them from all others, and their constant jerky motions make them very conspicuous. Bees announce their presence by a noisy humming. The beetles of the genus Calopteron have their wing-cases curiously distended, and move them up and down, so as to attract attention. . . . The reason in all these cases appears to be the same as Mr. Wallace has shown to hold good with banded, hairy, and brightly coloured caterpillars. These are distasteful to birds, and in consequence of their conspicuous colours, are easily known and avoided."

Belt goes on to make the general point that what holds true for the insect mimes is also true for the vertebrates. Among mammals, the foul-tasting skunk is colored bright black and white, very conspicuous in the crepuscular hours of dusk and dawn when skunks are most active, boldly warning nocturnal predators like owls and foxes to steer clear. Among reptiles, the beautifully banded coral snake, whose bite is deadly, is conspicuously marked with bright bands of black, yellow, and red. The coral snake, too, has imitators in the tropics—snakes that are not venomous but are banded black, yellow, and red in imitation.

Finally, Belt found a conspicuous frog and confirmed his ideas on the relation of conspicuous coloring to inedibility with a brief experiment: "In contrast with . . . obscurely coloured species, another little frog hops about in the daytime dressed in a bright livery of red and blue. He cannot be mistaken for any other, and his flaming vest and blue stockings show that he does not court concealment. He is very abundant in the damp woods, and I was convinced he was uneatable so soon as I made his acquaintance and saw the happy sense of security with which he hopped about. I took a few specimens home with me, and tried my fowls and ducks with them, but none would touch them. At last, by throwing down pieces of meat, for which there was great competition amongst them, I managed to entice a young duck into snatching up one of the little frogs. Instead of swallowing it, however, it instantly threw it out of its mouth, and went about jerking its head as if trying to throw off some unpleasant taste."

Belt does not identify the species of frog involved, but his description conforms to the "arrow-poison" frog, which natives of the Neotropics have long used to make the famous curare poison for their dart tips—the same poison Charles Waterton sought on his first journey into the forests of Guiana. To extract the poison from these frogs, the Indians hold them over fire until the poison oozes from their skin. The poison has a deadly effect on the heart and nervous system. One may wonder how the tropical American aborigines first discovered what it took the white man until 1870 to find out. Did they observe other animals rejecting the frogs as a food item, or did some Indian of the long ago have to advance the knowledge and hunting techniques of his people by eating the bright red and blue frog?

In all these observations of mimicry and conspicuous coloring, we learn several of the great lessons of tropical nature: that not all we meet in the world is as it seems, that much is deception and fraud. We learn that striking beauty often goes along with a poisonous personality. And finally, the most important lesson of all: be careful what you put in your mouth.

Moving westward, the central area, or Zone II, of Nicaragua is composed of grassy savannas. On these flatlands, in Belt's time as today, were grazed cattle herds, horses, and mules. "It is essentially a pasturage country," Belt writes. Eventually these plains enter a series of deep valleys cut by the many rivers flowing toward the Atlantic. The rivers in turn run down from the mountainous Sierra, cutting Nicaragua northwest to southeast like a spine, containing many volcanoes both dead and active.

Belt's main study in Zone II concerned a subject of lifelong interest to him: the Glacial Period or, as it is commonly called, the Ice Age. Riding down through the valleys of the Zone II foothills, he took particular note of the succession of rocks. He identified the bedrock as quartz and gneiss beds, which he had no doubt were the same rocks of the Laurentian formation that he had seen in Canada and Brazil. Belt called these "the very backbone of the continent, ribbing America from Patagonia to the Canadas."

Overlying these massive beds of quartz and gneiss were highly inclined and contorted schists—another igneous rock. But then along the river banks he spied exposed unstratified beds of gravel, deepening to a thickness of two hundred to three hundred feet on the undulating plains. These unstratified deposits consisted mostly of quartz sand with numerous large angular blocks and subangular boulders, some lying in the river beds measuring as much as fifteen feet across. Most of these boulders Belt identified as made of the same bedrock of quartz and gneiss, moved as far as eight miles from their parent rock.

What force could have driven such huge boulders so far? Belt thought boulders struck off Laurentian formations could only have been moved by the tremendous thrusting power of an ice glacier. The evidence of glacial action, he says, was as clear in the central zone of Nicaragua as in any Welsh highlands valley:

"I could no longer withstand the evidence that had been gradually accumulating of the presence of large glaciers in Central America during the glacial period, and these, once admitted, afforded me a solution of many phenomena that had before been inexplicable. The immense ridges of boulder clay between San Rafael and Yales, the long hog-backed hills near Tablason, the great transported boulders two leagues beyond Libertad on the Juigalpa road, and the scarcity of alluvial gold in the valleys of Santo Domingo, could all be easily explained in the supposition that the ice of the glacial period was not confined to extra-tropical lands, but in Central America covered all the higher ranges, and descended in great glaciers to at least as low as the line of country now standing at two thousand feet above the sea."

Having proposed that the Ice Age glaciers reached the American tropics, Belt next asked the logical questions: What became of the many genera of animals and plants peculiar to tropical America when a great part of the tropics was covered with ice and the climate of the lowlands was much colder than now? How was it that such peculiarly tropical groups were not exterminated by the cold of the glacial period? The answer Belt offered is both ingenious and fascinating:

"I believe the answer is, that there was much extermination during the glacial period, that many species and some genera, as for instance, the American horse, did not survive it, and that some of the great gaps that now exist in natural history were then made; but that a refuge was found for many species, on lands now below the ocean, that were uncovered by the lowering of the sea caused by the immense quantity of water that was locked up in frozen masses on the land."

Belt speculated that if the glaciers covered both hemispheres, the volume of water necessary for their construction must have lowered the general planetary level of the oceans by at least a thousand feet. Thus the animals and plants of tropical America would have migrated to these lowland margins, creating a great bio-refuge from the colder climate. The very accumulation of ice that made life impossible in the tropics during the glacial period provided an alternate habitat by draining the seas. (In my own home county of

Cape May, New Jersey, which is formed entirely of glacial moraine, fishermen find fossils of megafauna dating to the glacial period in their dredge nets up to 150 miles out to sea on the continental shelf.) Belt's idea was that the immense accumulation of ice over both poles reached through the temperate zones and encroached on the tropics into equatorial America, draining the seas. Lands now submerged would have been uncovered then. When the glaciers began to melt at the close of the Ice Age, the waters sweeping back into the inhabited lowlands must have caused calamitous floods of a truly biblical character. Belt wondered aloud if this could be the source of the legends of deluge we find in nearly all human mythologies.

Riding westward again, we come to the third zone of Nicaragua, skirting the Pacific—a country, Belt tells us, of fertile soils, where all the cultivated plants and fruits of the tropics thrive abundantly. As Belt traveled through this Pacific zone, he was puzzled at how the forest abruptly ended along an irregular line meeting grassy savannas. He saw that the soils and bedrock were of the same type on both forest and savanna. The contours of both habitats were similar. Nor was it lack of moisture: both forest and savanna in the Nicaraguan Pacific zone received equally at least six months of rain per year.

Why, then, should the dark promontory of trees abruptly segueway into the savannas? After almost five years of observing this phenomenon, Belt concluded that the forest had formerly extended much farther toward the Pacific. It was human perturbation that had transformed forested areas into savannas. "The ancient Indians of Nicaragua were an agricultural race," Belt wrote. "Their principal food then as now being maize." He described how the natives' traditional agricultural practices degraded the forests:

"They cut down patches of the forest and burnt it to plant their corn, as all along the edge of it they do still. The first time the forest is cut down, and the ground planted, the soil contains seeds of the forest trees, which, after the corn is gathered, spring up and regain possession of the ground, so that in twenty years, if such a spot is left alone, it will scarcely differ from the surrounding untouched forest. But it does not remain unmolested. After two or three years it is cut down again and a great change takes place. The soil does not now contain seeds of forest trees, and in their stead a great variety of weedy-looking shrubs, only found where the land has been cultivated, spring up. Grass, too, begins to get a hold on the ground."

Slash-and-burn agriculture had been going on since long before the Spanish Conquest. Belt distinguished brilliantly between those pre-Columbian peoples who grew maize and those tribes that depended on a hunting-gathering way of life:

"When Florida and Louisiana were first discovered, the native Indian tribes all cultivated maize as their staple food; and throughout Yucatan and Mexico, and all the western side of Central America, and through Peru to Chile, it was, and still is, the main sustenance of the Indians. The people that cultivated it were all more or less advanced in civilisation. . . . It is likely that these maize-eating peoples belonged to closely affiliated races. . . . From Cape Gracias a Dios southward, the eastern coast of America was peopled on its first discovery by much ruder tribes, who did not grow maize, but made bread from the roots of mandioca. . . . This fundamental difference in the food of the indigenes points to a great distinction between the peoples [of America]."

It is odd that Belt, who was perceptive enough to see the divide between the settled agricultural societies of maize eaters and the hunting-gathering societies of mandioca eaters, does not take the next step to show the two cultures in relation to their respective climates and geography. Advanced farming cultures based on maize grow on dry forest areas with definite rainy season–dry season patterns and relatively receptive soils. Mandioca eaters live in moist or rain forest areas where, despite seasonal variation, it is generally too wet to plant good corn crops, and the soil is relatively infertile. Thus one may find maize-growing Mayas on the fertile limestone lowlands of the Yucatan and in the mountains of Guatemala. But in the Mosquitia swamps or the Amazonian rain forest, one finds hunter-gatherers depending on mandioca roots. To put it simply, cultures, like species, adapt to their habitats.

Belt is apparently wrong, however, when he suggests that dry tropical forest areas converted to maize cultivation through slash-and-burn agriculture are repeatedly trashed to plant successive crops until the forest is permanently altered. Evidence has been found that seems to demonstrate instead that before the Spanish Conquest, traditional native American agriculture included a fallow and regenerative period, allowing the forest to grow back; when the conquistadors appeared on the shores of the New World, more than 90 percent of the territory was still clad in green forest. Without the help of manure or the boost of chemical fertilizers to improve his maize crops, the aboriginal farmer moved his *milpa* to newly burned terrain, allowing the forest to regenerate on former cropland. The same land might be reused, but not for several generations. Belt was essentially looking at the conditions of 1868 as if they had remained the same since 2500 B.C. Yet before the Spanish Conquest the population density was so low, and the yield of this amazing food source so high, that several thousand years could go by under the slash-and-burn regime with essentially no loss of tropical dry forest. The pre-Columbian system was essentially in balance with the ecosystem.

Belt remarks the interesting theory, probably correct,

that before the advent of the Europeans, the cornbelt of the tropics supported many more people than at present, at a better standard of living in nutritional terms—and, I would add, doing less damage to the environment. The maize-farming Mayan civilization reached a level of complexity and sophistication to equal the cultures of the Tigris-Euphrates Valley and Central Europe. But the weather was better in America.

In describing the "degeneration" of tropical American society since the Conquest, Belt wanders into a field he cannot, regrettably, read with the same accuracy he reads the rocks: "War is not always a curse. . . . Before the Spanish conquest no small isolated communities could exist. Those in which the tribal instinct was strongest, who stood shoulder to shoulder with their fellows, reverenced and obeyed their chiefs, and excelled in feats of strength and agility, would annihilate or subjugate the weaker and less warlike races. It was this constant struggle between the different tribes that weeded out the weak and indolent, and preserved the strong and enterprising; just as amongst many of the lower animals the stronger kill off the weaker, and the result is the improvement of the race, or at any rate the maintenance of the point of excellence at which it had arrived in former times.

"Since the Spanish conquest there has been no such process of selection in operation amongst the Indians. The most indolent can obtain enough food, whilst the climate makes clothing almost a superfluity. The idle and improvident live their natural terms of years, and increase their kind even faster than the provident and industrious. The tribal feeling is destroyed; the selfish and sensual instincts are developed, and year by year the Indian degenerates."

To suggest that war weeds out the weak and lazy and preserves the strong and enterprising, resulting in the improvement of the race, is a fine grade of Victorian myopia.

To go further and analyze social violence as a form of natural selection is simply bad science.

The problems Belt saw in Central America did not begin with the elimination of intertribal warfare after the Spanish Conquest but with the Spanish conquistadors effectively decapitating Central American societies, leaving them without their aristocracy, priesthood, and educated elite. The Spanish burned their books, set fire to the cities, and reduced the population to an illiterate peasantry. The Spanish also introduced European diseases such as smallpox; along with the inauguration of slavery and peonage, this reduced the native population by 90 percent in the first hundred years after the Conquest. Finally, the introduction of European plantation-style farming, monoculture, and particularly cattle ranching disrupted the traditional agricultural regime while beginning the devastation of tropical forests now reaching its sad climax. Dan Janzen has calculated that less than 1 percent of the tropical forest cover present at the time of the Conquest remains intact today.

Belt, like most nineteenth-century naturalists, felt duty-bound to mix his perceptive science with pseudoscientific ideas about his own species. He was no social philosopher. He is fine when he observes that hairy-bodied people would be at a disadvantage in the tropics for defense against pests and small biting critters. But all too soon he is wallowing in racist theorizing on the inferiority of Negroes and Indians to whites. Readers are treated to windy lectures on the lassitude of the Nicaraguan natives, and finally, to the perversion of science itself: social Darwinism, that absurd Victorian idea that because nature rewards the modification of species on the basis of survival, the powerful and strong therefore have a natural right to oppress the weak—and, by extension, mighty nations have the right to push around poor and vulnerable peoples. Could Belt's viewpoint have had anything to do with the fact that he worked for a British gold mining company, exploiting the natural resources of a politically weak and factionalized banana republic?

The modern reader is apt to forgive Thomas Belt the imperialist prejudices of his time in view of his fascinating revelations in tropical engineering. The difficulty, however, is that the ideas of Belt's time largely formed the common philosophy and values of our time—just as the biologists of our day stand on Darwin's shoulders. Harebrained ideas of social competition and racial superiority, mistakenly based on evolutionary theory as suggested by Belt and others, were taken up by less scientific men and retailed around to form the common cultural foundation of racial ideas widely accepted in the Americas today. Do we not live in a time when "only the strong survive"? Do young people not kill each other believing that "might makes right"? Belt's valuable contributions to our understanding of evolutionary design in the tropics must be distinguished from the half-baked notions about humans he sometimes purveyed as scientific truth.

7

Mr. Stephens and Mr. Catherwood

THEY WERE A FINE PAIR of nineteenth-century gentlemen. During all their travels, discoveries, and scrapes in Chiapas, Yucatán, and Central America, they never called each other anything but "Mr. Stephens" and "Mr. Catherwood." Like true Victorian explorers, both eventually died of malaria—but not before putting a scientific foundation under the infant field of American archaeology.

John Lloyd Stephens was as American as you can get. Born to a New Jersey merchant family that had fought on the winning side in the American Revolution, young Stephens was a Columbia University graduate, a successful lawyer in the New York of the 1830s with worldly ambitions—and the world at his feet. Active as a Jacksonian Democrat, Stephens was the pet speaker at the Tammany Hall Club, where New York City Democrats went to conspire against their political opponents. At the age of thirty-two, in 1837, Stephens wrote *Incidents of Travel in Egypt, Arabia Petraea, and the Holy Land*, an account of a year's tour of ancient monuments of the New East.

The 1830s were a golden time when Americans' boundless business optimism zoomed. Both frontier and urban growth contributed to a new way of viewing risks as opportunities. Americans had left the Revolution definitively behind but had not yet encountered the highly industrialized economy that would diminish the role of the individual in American culture. In the 1830s individual enterprise was still becoming enshrined as perhaps America's most cherished personal value, while salesmanship was becoming imbedded in the matrix of the American character.

Stephens's book was very much to the reading tastes of this newly forming business civilization: intelligent and well observed, but not gushy or visionary; refreshingly witty, energetic, and relentlessly positive. Not beholden to European literary conventions, Stephens wrote in plucky, confident style and had no fear of poking fun or of winking at a woman. His first book received rave reviews, including one by Edgar Allan Poe. Poe's review catapulted John Lloyd Stephens to a kind of literary and financial success that Poe himself never achieved. After a second volume on Greece, Turkey, and Russia proved equally popular in the bookstalls, Stephens, now a best-selling author, decided to explore Mexico and Central America.

Not a naturalist—in three volumes of some 300,000 words, scarcely an animal, mineral, or vegetable, is identified by name—Stephens was interested in discovering lost civilizations. He pulled strings in the State Department, then under Democratic control during the administration of

Frederic Catherwood's drawing of John Lloyd Stephens at Palenque, site of the ancient Mayan palace the two men discovered.

President Martin Van Buren, and wangled an appointment as American chargé d'affaires for Central America. The region was wracked by civil war in the aftermath of the collapse of Spain's American empire; diplomatic cover provided Stephens a measure of safety.

Frederic Catherwood was a British-born architect and landscape artist with little formal education. He could not have been more different from his future companion. An excessively formal Victorian gentleman with straight blond hair and wire-rim glasses, Catherwood was hardworking, taciturn, and by all odds the most gifted draftsman of his time. Unhappy in England, where he had committed the unpardonable sin of being born into the wrong social class, Catherwood departed for the Levant but wandered around the Mediterranean and the Middle East for ten years, thoroughly steeping himself in the artifacts of ancient civilizations. He drew the monuments of the Nile, Thebes, and Karnack to scale; excavated Egyptian ruins; painted famous panorama landscapes of Mt. Sinai and Baalbek; and eventually became professor of architecture at the University of Cairo as well as architectural adviser to Pasha Mehemet Ali, for whom Catherwood renovated some of Cairo's most ancient mosques. When it came to antiquities, few could speak with Frederic Catherwood's authority.

While living in New York, where he was making an excellent profit exhibiting his popular panoramas, Catherwood was introduced to Stephens by the owner of a bookstore they both frequented. It was Catherwood who brought to Stephens's attention a most curious book, the 1822 London edition of Captain Del Rio's *Description of the Ruins of an Ancient City Discovered Near Palenque in the Kingdom of Guatemala*. It had been published with wildly romantic illustrations by Count Waldeck, who had lived for two years at the jungle ruins.

Stephens borrowed the book and returned home. In

its pages he suddenly awakened, as he put it, to "the existence of ancient and unknown cities in America." He invited Catherwood to accompany him to Central America with the understanding that he would contribute illustrations of such monuments as they might find for Stephens's book. Catherwood was a gentleman of few words. "Yes," he replied, and together the two men set out for the meso-American tropics in the autumn of 1839.

Of course there had been much anecdotal evidence of ancient cities in the tropical jungles of the New World, dating back to the first moments of the Spanish Conquest. Diaz del Castillo, who was with Cortés in 1504, and lived to write *The Conquest of Mexico*, said the ruined cities he had seen with his own eyes rivaled Seville in size and beauty.

Humboldt was the first non-Spaniard of a scientific bent to travel extensively in Mexico, where he spent a year preparing a plan of reform for that country's silver mining industry. He personally visited the great pyramid temple at Cholula and collected drawings and information everywhere about archaeological remains. In the *Travels*, Humboldt mentioned the name of a lost city called Palenque. But he had not traveled in Yucatán, Chiapas, or the highlands of Guatemala, and so really had not scratched the surface. Even the great Humboldt had little idea that throughout the tablelands of Mexico and the adjacent Central American highlands, an area of approximately seventy thousand square miles, stood the stone relics of some sixty or more highly developed Mayan city-states, some dating back more than a millennium before the Conquest.

The words "Maya" and "Aztec" did not then appear in any dictionaries or encyclopedias. A blanket of the white man's severest oblivion was spread over America's pre-Columbian civilizations. Noble or otherwise, the Indians were universally considered savages. No one thought them capable of the technical sophistication and social organization necessary to engineer large-scale construction, neither buildings nor irrigation systems. If such cities existed, then, who had built them?

In 1834 Lord Kingsborough published his nine-volume work *Antiquities of Mexico*, proving "scientifically," as he said, that the people who had left monumental ruins behind were Hebrews descended from the Lost Tribes of Israel. Count Waldeck also drew the human figures from the sculptures on the ruins he visited as looking quite Semitic; but Waldeck differed slightly with his patron Kingsborough. Waldeck thought the Mexican ruins were built by the Egyptians.

This was the state of American archaeology when Stephens and Catherwood took the field in May 1839. The ruins of Palenque were not far into the jungle, only about eight miles from the town of Santo Domingo del Palenque—*palenque* meaning "palisaded" in Spanish. Setting out in military fashion at daybreak, Stephens and Catherwood were joined by another American, Henry Pawling, who had a brace of pistols and a double-barreled shotgun, in case anyone became unfriendly. They had actually been welcomed rather heartily in the town of Palenque, and no wonder: they bought out the stores for provisions, rented all the mules, and hired a brigade of local Indians to carry the gear on their backs, supported at the forehead with tumplines, in the time-honored Mayan tradition of human cargo transport.

Within two hours they had left the road and turned sharply uphill into an emerald rain forest. It was the beginning of the rainy season, and there had been deluges the night before. The little Rio Michol was overflowing its stone bed. Instead of a path they found beautiful small waterfalls cascading everywhere down the mountainside. The wet forest understory, bowing low under the weight of moisture, soaked the travelers as they crawled through it. The mules got stuck in the gullies. Stephens dismounted, rolled up his pants over his boots, and forged ahead, trailing blue smoke from his morning cigar. In that moment, he later wrote, he felt "nothing could speak so forcibly of the world's mutations" as this immense and silent rain forest, shrouding what he presumed once must have been a great avenue, thronged with people, leading to a great city. The "world's mutations" was a nineteenth-century buzzword having to do with the romantic idea of fate or destiny—things out of our hands, like the rise and fall of civilizations. This was Stephens at his sentimental worst.

Catherwood, meanwhile, cleared the steam from his spectacles and marched on. He heard the howler monkeys jeering down from the crowns of the mammoth mahogany trees. But he kept his eyes on the ground. There were masses of white stones in the forest. Limestone. Most of them were shaped for building blocks, twelve inches by twelve inches, but there were larger stones as well, and some were sculptured. Catherwood's pulse quickened. It could only mean one thing: they were poised on the discovery not just of some buried treasure but of an entire unknown civilization, buried in America's tropical jungles.

They spurred the mules up a steep ascent to a terrace covered with vegetation. Before they could make out its form, the Indians cried: "El Palacio!" The Palace! Stephens wrote: "And through the openings in the trees we saw the front of a large building richly ornamented with stuccoed figures on the pilasters, curious and elegant . . . in style and effect unique, extraordinary, and mournfully beautiful."

It was the building known ever after as the Palace at Palenque, a monumental longhouse complex set on top of a great stone pyramid. The Mayas usually erected these limestone block pyramids with rectangular temples at their summits, but they also raised many longhouses—presumably royal residences—all faced like the temples with brilliantly ornamented and painted stucco. The Mayas cleared and terraced enormous areas of the bush to build ceremonial ball courts and cemented plazas, where they erected stone monoliths carved head to foot with hieroglyphs. So much attention was later paid to the puzzle of deciphering the Mayan hieroglyphs that we have often underemphasized the remarkable engineering and architectural achievements of these tropical Americans.

All the great Mayan "cities" were actually ceremonial centers at the heart of much larger centers of population and agronomy. As Bishop de Landa, the destroyer of Mayan texts and first anthropologist of the Mayas, wrote, "In the middle of the town were the temples, with handsome squares, and around the temples were the houses of the rulers and the priests, and after these came those of the principal people."

The raised temples were the spiritual centers of the Mayas' world. Up these stairways to heaven climbed the Mayan royals and their priests to perform the sacrifices and rituals that kept the right deities involved in human affairs on a favorable basis: agriculture, birth, death, marriage, warfare, trade, foreign relations, and especially the accession and succession of kings. In the bas reliefs cut with awesome skill by sculptors wielding flint chisels, the Mayan dynasties told their stories and laid their claim to divine right for all the people gathered in the plazas to behold.

Stephens and Catherwood each fired a salute with Pawling's gun in celebration, then set up their camp in the front portico of the palace. Catherwood commemorated the scene in a famous drawing which shows the two gentlemen sitting at a stone table in the front corridor of the palace in coats and hats while the jungle vines snake their way down an enormous pilaster above their heads. That night they sat down to their first meal in a mood of elation, but in the middle of dinner a sudden downpour sent them scrambling for cover (the palace no longer had a roof). The pummeling rain continued right through the night. In the morning, sleepless and soaked to the bone, the explorers discovered that the water had drenched and spoiled their provisions. The tortillas had become a wet lump of cornmeal. None of the white men knew how to transform the lump back into tortillas. The Indians had refused to spend the night at the ruins, not out of fear or rebellion, but simply because they had to go to work at their *milpas*, or maize plots, at the break of day, according to Mayan custom. They understood, as Stephens and Catherwood apparently did not, what the rainstorm meant: it was the beginning of the rainy season, which is also the season of planting corn. And nothing was more important to the Mayas than corn, the basis of their existence.

The weather continued wet, and with the rains came the mosquitoes, "beyond all endurance," as Stephens put it. The explorers were prepared for hardship and even danger. They were prepared to defend themselves against jaguars and *banditos*. But nothing could have prepared them for the onslaught of the mosquitoes and the sleep deprivation the insects caused. Stephens wrote, "The slightest part of the body, the tip end of a finger exposed, was bitten. With our heads covered the heat was suffocating, and in the morning our faces were all in blotches."

They had no idea that mosquitoes carry malaria. Both men were infected and suffered bouts of fever throughout the rest of their lives. They were plagued as well by ticks and infested with lice. Of particular annoyance were what the locals called *niguas*, probably chiggers, which burrowed under the skin, "then laid their nits therewithin," so that the explorer became host not only to the adult but to the children as well. Stephens's foot became infected and swelled to the point where he could not wear a shoe. Pawling finally dug out Stephens's niguas with a knife, but the infection only grew worse. Catherwood was soon wracked with chills and fever, fever and chills. After sleepless nights of torture by the mosquitoes, Catherwood would fall asleep slumped over his easel during the day. Still, they got down to work.

For ten days they surveyed, measured, mapped, and drew. In one of Catherwood's most charming lithographs, he and a coatless John Stephens are portrayed in miniature, measuring the front of a stone longhouse while Pawling chases after a turkey in the nearby jungle with his gun. Stephens organized the excavations in his appealing role as the enterprising amateur. "My business," he tells us, "was to prepare the different objects for Mr. Catherwood to draw."

This meant directing the hired help—i.e., Pawling and the few natives he could get on a cash-and-carry basis—to chop down the trees and thick vines, and clear away the rubble and dirt, while Stephens stood by smoking his cheroots. The clearing did not go well at all, since the Indians had no axes, only their machetes. But as soon as he heard a shovel or hoe clang against stone, Stephens jumped to it and took over the removal of the last dirt from whatever was underneath. After all, it was his expedition, and he had the right of discovery.

Stephens also supervised the scrubbing and cleaning of the stones, a monumental job. All the time he lusted after those stones. His scheme was to buy up all the ruins he could, ship a collection of the ancient monuments back to New York, and launch a new American Museum of Antiquities. He would make a fortune as the American Lord Elgin.

By the time the Stephens-Catherwood expedition reached Palenque, Stephens could already boast that he had bought the entire ruins of Copán in Honduras for fifty dollars and was still negotiating the purchase of Quiriguá. He would try to buy Palenque, too, but was thwarted by Mexican law, which allowed only those foreigners who were married to Mexicans to own Mexican property. For a while Stephens considered marrying a young widow from the town of Palenque, but it never happened (he died a bachelor). Perhaps it was not completely by coincidence that P. T. Barnum would base one of his infamous hoaxes, the public exhibition of the so-called Aztec Children, on John Lloyd Stephens's description of the Mayan ruins. There was a good deal of the impresario, and more than a bit of the huckster, in Stephens.

Whether Catherwood disapproved of Stephens's wheeling and dealing we will never know. What we do know is that more than anything, Stephens understood only too well Catherwood's value to his own prospects, and he managed everything with a keen eye toward keeping his partner in business.

Catherwood's brilliant landscape of the Palace at Palenque shows the palace from a modest distance, looking north, and slightly rotated, so that one sees past it to the surrounding lowland plains of Chiapas and Tabasco. In order to gain that perspective, Catherwood had to climb a steep hill and clear a considerable amount of bush. The entire foreground of the landscape is strewn with the shrubs and vines. But the perspective is perfect, because it provides an accurate idea of the immensity of the mound on which the palace is perched. The Mayas were the unsurpassed mound-builders of pre-Columbian America. They had no shortage of labor or materials. They brought dirt and stone rubble to the site in baskets on the backs of the commoners. They heaped it as high as they could, with still enough room on top for the building. Then they faced the mound with terraced limestone, which they cut from numerous quarries with broad flint chisels and apparently hauled to the site on sledges pulled by men. Their environment produced no quadrupeds large enough to haul carts, so the Mayas had never faced the need for inventing the wheel. Still, without pulleys, it is difficult to see how they hoisted the larger stones to the top of the mound.

Only after they had constructed the mound were the Mayas ready to build the structure on top. The Palace at Palenque is unique in having a three-story tower rising above the south end. As a result, in Catherwood's landscape the palace looks alive, like a lioness sprawled regally upon her pedestal with her lithe neck stretched up full, overlooking the jungle. By lightly sketching in the plains and hills all the way to the horizon, Catherwood shows how this one building dominates a vast lowland plain.

The limestone and cement structure itself is 228 feet long by 180 feet deep, and 25 feet in height from the terrace. The terraced stones form a base 75 or 100 feet high. The front contains 14 doorways, each 9 feet across. Virtually the entire façade of the palace was once ornamented with giant stucco reliefs of gods, plants, animals, and human figures with flattened heads wearing plumed headdresses. The reliefs were cut with fine flint chisels. They are each 6 by 10 feet and richly bordered, often with a single cartouche of hieroglyphs. Stephens confessed a complete inability to understand these strange figures, but he could only admire the work. "The stucco is of admirable consistency, and hard as stone," he wrote. "It was painted, and in different places about it we discovered the remains of red, blue, yellow, black and white." Later evidence has shown that the Mayas sometimes painted their pyramids red, scarlet, or purple. This must have made a dramatic contrast with the multihued greens of the tropical forest.

Within the palace the explorers found two extensive rectangular courtyards filled with rubble and thickly overgrown with trees. An excavating operation revealed a set of thirty-foot-long steps, on each side of which were propped huge stone bas reliefs showing "strange grim and gigantic figures."

At first Stephens and Catherwood thought the palace was the only building on the site. "From the Palace," wrote Stephens, "no other building is visible." The forest cover was so heavy that the explorers could not even see that at the very foot of the southwest corner of the palace rose an impressive 110-foot pyramid-mound. Not until Stephens climbed the ruins of the palace's 30-foot tower did he discover another building a stone's throw away.

This structure, too, was obscured by dirt and vegetation. Stephens called it simply Casa de Piedras Numero Uno (House of Stones Number One). We know it today as the Temple of the Inscriptions. Plants grew from the roof; the tenacious roots of the trees had dislodged the stone stairs of the pyramid. The ascent was so steep and treacherous, Stephens tells us, that if the first man dislodged a stone it would bound down on the heads of those behind. They had to hoist themselves up by grabbing roots and branches.

The front of the temple faced north toward the plaza and was entered by five grand doors; but the relief work was badly eroded. The faces of the figures were gone, though there were still bits of paint—bright red, black, yellow, scarlet, purple, and white. The inky darkness of the interior was home to fruit bats, lizards, spiders, and tarantulas.

They crept forward inside the room. The Mayas did not know the true arch, and built their roofs by what is called the false arch method, i.e., terracing the stones from each side until they got close enough to cover the opening with a capstone. This method limited the size of Mayan temples, which therefore tend to be cramped, but because only the king, priests, and sacrificial victims had to go inside the temples, it was not a significant hindrance.

Stephens's local guide Juan was leading the way with a pine torch when he let out a startled cry. Under the dank moss of the side walls appeared two huge flat, limestone tablets covered with hieroglyphs. Stephens ordered that brushes and water be carried up. As the cleaning revealed their form, Stephens and Catherwood were swept by wonder. The precise calligraphy, the mathematical regularity, but above all the human figures, the faces of the kings of the jungle, shook Stephens and Catherwood to their boots. "The impression made upon our minds by these speaking but unintelligible tablets I shall not attempt to describe," Stephens wrote.

Catherwood had brought with him a camera lucida with which to obtain the most accurate visual renderings technically possible at that time. Only twenty years after the Stephens-Catherwood expedition, Mathew Brady was documenting the American Civil War by daguerreotype. And soon thereafter photographer-archaeologists like A. P. Maudsley were recording unexcavated Mayan temple mounds on black-and-white photographic film. The camera lucida was an eighteenth-century device for aiding landscape artists. A plain glass prism projected a virtual image of an object onto the surface of a sheet of paper placed beneath the glass, allowing the object's outline to be traced with the utmost precision. Developed for drawing accurate perspective, it turned out to be a worthy tool for Catherwood in tracing the tablets.

The explorers lit the vault with torches and candles, and Catherwood, setting up the camera lucida, began the exacting labor of tracing the pebble-shaped glyphs line by line. At Copán his expert eye had quickly surmised that the hieroglyphs bore no resemblance whatsoever to cuneiform, Sanskrit, or any written language he had encountered. It was not hard to see that the hieroglyphs at Palenque were generally the same as at Copán. The implication was clear. Stephens wrote, "There is room for the belief that the whole of this country was once occupied by the same race, speaking the same language, or at least, having the same written characters."

Although John Lloyd Stephens is commonly called the father of American archaeology, archaeology as a science really begins with Frederic Catherwood's precise drawings of the stone tablets and monoliths in Central America. By accurately rendering these stones, the process of scholarship leading to decipherment began. And although no key akin to the Rosetta stone has ever been found, 150 years of avid scholarly debate has revealed the Mayan calendar and dating systems, the names of the gods, kings, priests, and nobles, their wives, and children. By now the histories and mythologies fashioned by the jungle dynasties that built centers such as Palenque have been largely decoded. It is interesting that none seems to say anything definitive about the Mayan collapse in the ninth century A.D., when the ceremonial centers were abandoned. That remains the enduring mystery of these great mound-builders.

Having finished copying the inscriptions, Catherwood and Stephens crossed the small *quebrada* flowing along the rear or western side of the palace and came upon three more temple mounds framing another plaza. They are now known as the Temple of the Cross, the Temple of the Foliated Cross, and the Temple of the Sun, called by archaeologist Michael Coe "the most perfect of all Maya buildings," with its elegant proportions, mansard roof, and fantastic forty-two-foot-high roof comb.

Catherwood made accurate architectural drawings of these stately temples. His front elevations provided what Stephens called "proof against the many irrational speculations" regarding the origin of meso-American ruins—that they were constructed by people of another continent. "The striking feature of these ruins is that the buildings stand on lofty artificial elevations," Stephens wrote. The Egyptian pyramids, on the other hand, are uniform and complete structures unto themselves. To the Egyptians, the pyramidal form is no more than an engineering solution to the problem of building high structures on a solid base. It never occurred to the pharaohs or their architects that pyramids should act as foundations for a structure on top. The pharaohs never intended to perform public ceremonies atop their pyramids.

For Stephens, the architectural evidence was persuasive. The relics of Palenque and the forty-four other sites he visited with Catherwood were like nothing built in Europe, Asia, China, Africa, or the Near East. Although some Mayan monoliths weighed as much as thirty tons, their standard twelve-by-twelve limestone construction block was much smaller than the ancient engineers of the East had employed. The Egyptians had the wheel and pulley and could use blocks a hundred times larger; and in the Old World builders could harness the locomotive power of draft animals.

Stephens concluded that the construction at Palenque was distinct, original, and native. "What we had before our eyes was grand, curious, and remarkable enough," he wrote. "Here were the remains of a cultivated, polished, and peculiar people, who had passed through all the stages incident to the rise and fall of nations; reached their golden age and perished."

But had the cultivated and peculiar people who built Palenque and the other meso-American ceremonial centers in fact perished?

Catherwood worked as if in a trance, determined to finish his visual record of Palenque despite the fever, the rains, parasites, and exhaustion. His drawings fall into three general categories: the architectural elevations; the general scenes or landscapes; and the copies of the tablets, which often included central images surrounded by glyphs. Catherwood's masterpiece from Palenque comes from the back wall of the altar of the Temple of the Sun, a tablet which tells the story of King Chan Bahlum receiving the scepter from his dead father King Pacal, whose tomb lay deep within the Temple of the Inscriptions. It was one of the last copies Catherwood accomplished at Palenque. Little did he know that these tablets constituted the most complete surviving record of any Mayan dynasty. After twenty-eight days and nights of labor, with little sleep and little food, Catherwood crumpled. Under lashing rains and whipping winds, Stephens dragged away a feeble Catherwood. They quit the ruins of Palenque on June 1, 1840.

One short but profound intellectual leap remained for Stephens. Having recognized that the architecture, literacy, and artistry at Palenque and the other sites they visited could only have been made by an indigenous race, it remained for Stephens to name it. He had experienced first-hand the destructiveness of the climate and witnessed the rank growth of the tropics, and he was not inclined to believe that a single stone could have survived the passing of two or three thousand years. The fairly intact condition of the ruins—as well as the survival of some hardwood beams at Uxmal—led Stephens to think the ruins dated from a more recent time.

In the Dresden and Vienna codices published in Humboldt's works, Stephens saw the only two meso-American manuscripts to escape the tragic destruction of the Conquest. Although recognizing differences between carving in stone and writing on agave paper with ink dyes, Stephens could not fail to observe that the literate artifacts of the natives at the time of the Conquest and the undeciphered glyphs at Palenque and Copán were essentially the same. "We are not warranted in going back to any ancient nation of the Old World for the builders of these cities," Stephens said. "They are not the work of people who have passed away and whose history is lost, but . . . creations of the same races who inhabited the country at the time of the Spanish conquest, or some not very distant progenitors."

If only a few centuries stood between the existence of the great cities and the advent of Cortés and his men, then the descendants of the former had not perished, or even vanished, but, veiled and unseen, had scattered in hidden jungle hamlets and retired in silent refuges, continuing stubbornly and in secret to carry on the sacred traditions that gave meaning and order to their lives. With the publication in 1841 of John Lloyd Stephens's *Incidents of Travel in Central America, Chiapas, and Yucatán*, the Mayas may be said to have returned to history.

The idea that America had its own indigenous civilizations whose ancient accomplishments rivaled anything found in the ancient Old World captured the imagination of a generation of Americans just beginning to struggle for an identity distinct from Europe. With his populist temperament, Stephens saw to it that his book was published at a price the general public could afford. Nor did he stop there.

In the *Incidents of Travel*, Stephens vociferously protested reports of "the immense labour and expense of exploring these ruins" as "exaggerated and untrue." The idea that only governments or institutes could afford such ventures gave Stephens fits. Eight or ten young "pioneers having a spirit of enterprise equal to their bone and muscle," as he put it, could lay bare a city like Palenque in a few months' time. "Any man who has ever cleared a hundred acres of land is competent to undertake it," he insisted.

The United States had plenty of such men in 1840. Stephens remarked it would cost less for a young American to travel in Central America for a few months than to spend the winter in Paris. And it would be more worthwhile. Stephens was the great American traveler as everyman, the amateur spirit incarnate of a democratic archaeological science.

Unfortunately the United States was already stretching toward Manifest Destiny. Within a decade Americans would invade Mexico not with explorers but with marines. An indication of how rapidly public excitement about the Mayan discoveries faded was that after the phenomenal success of Stephens's books, Catherwood could not raise sufficient money to publish his lithographs of the Mayan ruins.

Catherwood's personal fortunes did not fare much better. His panoramas burned in a New York City fire, and he was obliged to change professions. He became an engineer and built the first railroad in British Guiana. Stephens went into the steamship and railroad business and became the main American promoter for the building of the Panama Canal. He died at forty-six, but not before becoming president of the Panama Rail Road Company and hiring none other than Frederic Catherwood as his chief engineer.

8

W. H. Hudson and the War Against Nature

SOMEONE OF HIS OWN GENERATION once said that if W. H. Hudson could have molded the world to his liking, it would have come out as one vast bird sanctuary. This only demonstrates how completely this great naturalist and writer was misunderstood in his time.

Hudson was born on the Argentine pampas in 1841, at about the same time that Thoreau was hewing his hermit's hut at Walden Pond. He died in London in 1922, eight years after Europe plunged the world into World War I. His life spanned, then, the most rapid and rapacious period of human alteration of the planet's environment in all history until that time, accomplished by a burgeoning industrialism and imperialism, fueled by a brash belief in progress. By the end of Hudson's life in the 1920s, the relentless hunger for raw materials and agricultural production had left virtually no corner of the world unmolested. With the closing of the New World frontier, the replacement of many species in varied habitats by fewer, domesticated species in simplified habitats became the pastoral emblem of a century of economic development.

Hudson described the process as it took place in the valley of the Rio Negro in his book *Idle Days in Patagonia*: "The valley soil is thin, being principally sand and gravel, with a slight admixture of vegetable mold; and its original vegetation was made up of coarse perennial grasses, herbaceous shrubs and rushes: the domestic cattle introduced by the white settlers destroyed these slow-growing grasses and plants, and, as happened in most temperate regions of the globe colonized by Europeans, the sweet, quick-growing, short-lived grasses and clovers of the Old World sprang up and occupied the soil. Here [in Patagonia], however, owing to the excessive dryness of the climate, and the violence of the winds that prevail in summer, the new imported vegetation has proved but a sorry substitute for the old and vanished. It does not grow large enough to retain the scanty moisture, it is too short-lived, and the frail quickly perishing rootlets do not bind the earth together, like the tough fibrous blanket formed by the old grasses. The heat burns it to dust and ashes, the wind blows it away, blade and root, and the surface soil with it, in many places disclosing the yellow underlying sand with all that was buried in it of old."

The Western war against nature, always previously fought on even terms, was effectually won in the Temperate Zone with the creation of the dry plains and desert. Yet Hudson had known an immense, wild, and free savanna of the American hemisphere before Darwin published *Origin of the Species*. And he knew it not as some visiting European, self-consciously absorbing facts, nor as some urbanized do-gooder, discovering the benefits of wilderness from the safe havens of privilege, but simply as a boy knows his own neighborhood. From the age of six he was given a pony and allowed to ride out from his father's frontier *estancia*, a few miles from Río de la Plata, to wherever his spirit took him, staying out as long as he liked. He was a tall, strong, athletic boy, a good rider and a good shot, and from that early time he displayed the essentials of the gifted naturalist—the stamina and curiosity to roam about out of doors and observe; an emotional identification with animals and landscapes; and, most of all, the frank delight and sense of freedom in being alone in a vast natural solitude.

The great area of level plains of Argentina called by English writers the pampas, and by the Spanish more appropriately *la pampa* (from the native Quechua word meaning open space or open country), comprises about 200,000 square miles of humid, grassy steppelands. The pampas extend roughly from the Atlantic Ocean halfway to the Andes, and from the area of the Río Paraná, in latitude 32 degrees, south into the sandy, more or less barren district known as the sterile pampa, which in turn passes gradually into the desiccated plains of Patagonia. Although the pampean country had been colonized by Europeans since the middle of the sixteenth century, immigration was on too limited a scale to effect great changes. Until the Argentine government launched a determined campaign to rid the country of aborigines in 1879, the Indians were able to keep the invaders from most of their ancestral hunting grounds.

In Hudson's youth the pampas teemed with birds, snakes, insects, and a full complement of South American mammals—viscachas, coypus, hares, cavies, foxes, wolves, skunks, weasels, opossums, and four different kinds of armadillos, descendants of those edentate megafauna whose fossilized remains Darwin enthusiastically collected in

No better book on the ornithology of "the great bird continent," as he liked to call South America, has appeared since W. H. Hudson's *Birds of La Plata*.

Patagonia. A terrestrial sea of tall, waving, monochrome grasses stretched away from the gates of Buenos Aires for hundreds of miles, broken only infrequently by man-made atolls of trees planted round scattered ranchos, and by broad lagoons and marshes along the Plata and Paraná rivers. Over these vast unfenced plains roamed an incredible multitude of wild, domesticated, feral, and semidomesticated cattle, horses, sheep, pigs, and guanacos, or llamas, a New World relative of the camel. The domesticated herds were tended by equally wild, or only half-domesticated, gauchos, the Spanish-American equivalent of America's Western cowboy. Jaguars and pumas were among the chief predators of the glades and pastures along the Plata and Paraná. The lofty rhea, or American ostrich, was a common sight, racing with its odd habit of splaying one wing up "like a great sail."

The hot southwesterly wind of the interior, known as the *pampero*, heralded at times by thick storms of dragonflies, raced across the pampas with cyclone force, and sudden hailstorms literally knocked herds over dead. The wind was seldom at rest on those extensive level plains, wrote Hudson, giving out "an endless variety of sorrowful sounds, from the sharp fitful sibilations of the dry wiry grasses on the barren places, to the long mysterious moans that swell and die in the tall polished rushes of the marsh."

In those days a ride of two hundred miles from Buenos Aires was enough to bring one well west of the farthest frontier outpost, into a purple land displaying "those perfect sunsets seen only in the wilderness, where no lines of house or hedge mar the enchanting disorder of nature, and the earth and sky tints are in harmony." There the young Hudson would lie back with his head on his horse's rump, galloping for hours across "the matchless grass, which spread away for miles on every side, the myriads of white spears, touched with varied color, blending in the distance and appearing almost like the surface of a cloud."

More affecting than even the sheer space or the grand diversity of animal life on the pampas, the pantheon of animistic gods had not as yet been tossed off the ranch. Hudson sensed their presence early. Lying in bed listening to the night-migrating flocks of plovers overhead, he was inducted into that "enchanted realm, a nature at once natural and supernatural." Then he would go out under the full moon and listen to the birds, the trees, the whispering grasses, experiencing something "similar to the feeling a person would have if visited by a supernatural being."

Hudson was a modest man of gorgeous literary gifts, and he found it awkward to write later of this profoundly spiritual aspect of his boyhood. He was well aware that the orthodox Western intellect, at its worst, derides pantheism as primitive and irrational superstition, or at best makes the worship of nature a subbranch of art criticism, as though the apprehension of natural powers is a proper field only for artists dealing primarily with beauty. When Hudson was finally moved to write the autobiography of his formative years on the pampas, it was only because of a lucid vision of his boyhood, brought on by a near-fatal fever. This vision, however, was so powerful, detailed, and long-lasting that Hudson felt more compelled than inspired to write down the memories of his boyhood on the pampas.

When he got to the point in his story where he could no longer mask or ignore his animistic experiences, he attempted to explain his spiritual sense of nature in the same patient, precise, reasonable, modulated tones he used to discuss each aspect of natural history. Animism, Hudson said cautiously, was "that sense of something in nature which to the enlightened or civilized man is not there, and in the civilized man's child . . . is but a faint survival of a phase of the primitive mind." He did not mean to expound, he said, "the theory of a soul in nature, but the tendency or impulse or instinct, in which all myth originates, to *animate* all things; the projection of ourselves into nature; the sense and apprehension of an intelligence like our own but more powerful in all visible things."

Hudson thought this impulse probably existed to some degree in all humans, but "withers and dies" early in life among those raised in large towns and cities, "where nature has been tamed until it appears part of man's work." He thought animism persisted and lived in many "born and bred amidst rural surroundings, where there are hills and woods and rocks and streams and waterfalls, these being the conditions which are most favourable to it." In his own case, writing in England at the age of seventy-six, five years before his death, Hudson confessed that his own animism had never been wholly outlived. "This same primitive faculty which manifested itself in my early boyhood, still persists, and in those early years was so powerful that I am almost afraid to say how deeply I was moved by it."

Animism, pantheism, the sense of a holy intelligence in the natural order is so central to understanding W. H. Hudson that it is worth dwelling on his own exploration of the subject for a moment longer. At first, Hudson thought, he was "unconscious of any such element in nature." The childish delight he experienced in all natural things was "purely physical." In one of his many memorable passages, Hudson recounts this earliest, preanimist phase:

"I rejoiced in colours, scents, sounds, in taste and touch, the blue of the sky, the verdure of the earth, the sparkle of sunlight on water, the taste of milk, of fruit, of honey, the smell of dry or moist soil, of wind and rain, of herbs and flowers; the mere feel of a blade of grass made me happy; and there were certain sounds and perfumes, and above all certain colours in flowers . . . which intoxicated me with delight. When, riding on the plain, I discovered a patch of scarlet verbenas in full bloom, with a moist, green sward sprinkled abundantly with the shining flowerbosses, I would throw myself from my pony with a cry of joy to lie on the turf among them and feast my sight on their brilliant colour."

About his eighth year Hudson became distinctly conscious of something more than his childish delight. He would experience a new kind of thrill, which he described as sometimes pleasurable but at other times startling, or even so poignant as to become frightening. He wrote, "The sight of a magnificent sunset was sometimes almost more than I could endure, and made me wish to hide myself away." Certain flowers could set off these powerful feelings in him. Birds could have a similar effect. But this new, mysterious sensation was evoked more powerfully by trees than by anything else:

"I used to steal out of the house alone when the moon was at its full to stand, silent and motionless, near some group of large trees, gazing at the dusky green foliage silvered by the beams; and at such times the sense of mystery would grow until a sensation of delight would change to fear, and the fear increase until it was no longer to be borne, and I would hastily escape to recover the sense of reality and safety indoors, where there was light and company. Yet on the very next night I would steal out again and go to the spot where the effect was strongest, which was usually among the large locust or white acacia trees, which gave the name of Las Acacias to our place. The loose feathery foliage on moonlit nights had a peculiar hoary aspect that made this tree seem more intensely alive than others, more conscious of my presence and watchful of me."

There is a subtle, little explored, connection between geography and religious experience. There is a religion condign to desert nomads, and to fishermen on the shores, and to riders on the plains. Jesus says in the Gospel According to Thomas: "The kingdom of the Father is spread upon the earth, and men do not see it." Animism, however, unlike the universal, monotheistic religions, is always a localist phenomenon, with its glow emanating from particular places and the spirits that haunt them. The animistic gods of Kilimanjaro can mean but little to someone from Dubuque. The power of Fuji can mean nothing to someone from Kilimanjaro. Almost certainly this is the reason that whatever flotsam of pantheism remains afloat in the great global sea of consumerism is found among remote, rural, stationary cultures. Hudson recognized as much in a narrow and winding but deep essay innocently titled "Aspects of the Valley," in which he contemplates the meaning of the Rio Negro to the aboriginal inhabitants of that river's valley.

Dwelling from generation to generation by the river's banks, the natives would come to know that "sinuous shiny line" of water as "the one great central unforgettable fact in nature and man's life"—in short, there chief god manifested. Hudson went on:

"The shining stream was always in sight, and when, turning their backs on it, they climbed out of the valley, they saw only grey desolation—a desert where life was impossible to man—fading into the blue haze of the horizon; and there was nothing beyond it. On that grey strip, on the borders of the unknown beyond, they could search for tortoises, and hunt a few wild animals, and gather a few wild fruits, and hard woods and spines for weapons; and then return to the river, as children go back to their mother. All things were reflected in its waters, the infinite blue sky, the clouds and heavenly bodies; the trees and tall herbage on its banks, and their dark faces; and just as they were mirrored in it, so its current was mirrored in their minds."

But when nature is no longer the dominant presence, her spell is broken. And when one abandons that place, nature's power necessarily fades. The pantheistic gods and goddesses travel no better than a bologna sandwich on a hot day. Nonetheless Hudson was never, in his later life in England, able completely to swallow the chilly materialist dogma of natural selection. He remained something of a shy mystic, though a sane, skeptical, and eminently reasonable one, always first to call his own animism "primitive" and "childish."

It is true, but misses the point, that Hudson spent a good deal of time and energy in his old age writing blistering pamphlets against the slaughter of birds for ladies' millinery, and for the liberation of all caged birds. The birds never had a better friend and champion than W. H. Hudson. But the nostalgia and sentimentality implied in the notion of a life dedicated to creating bird sanctuaries—sort of concentration camps for the losing species after humans had mercilessly blitzkrieged nature—would have outraged Hudson, who once described England as "a glorified poultry farm." It misses the central point of Hudson's spirited rebellion against our usurpation of nature, our wanton destruction of land and life, our doomed attempts to control the very processes of universal change and set ourselves up not as the faithful flock of our own God but as gods ourselves, replacing the pantheon of nature. What Hudson wanted was not island refuges from industrial society but a recreation of the Argentine pampas of his boyhood, nothing less. And that is just what he used literature to do.

From about the age of fifteen, when sudden attacks of typhus and rheumatic fever nearly killed him, Hudson became a voracious reader. It was then he discovered Gilbert White's *Natural History of Selbourne*, that great classic of eighteenth-century naturalist literature in which the blessed English country parson undertook a complete natural history of one district, based upon his years of walks and observations in the meadows and woods of home. Few who have tasted the homey delights of *The Natural History of Selbourne* go away unaffected, and Hudson conceived the idea of making his native territory of La Plata his own "parish of Selbourne." As he recovered his health, he began riding farther and farther, staying out in the field for weeks at a time, keeping meticulous journals of his observations. Without time as a limiting factor, he patiently recorded the most intimate habits of the birds—"Never was there a better describer of the habits of birds," wrote Alfred Wallace when he reviewed Hudson's *Birds of La Plata* for *Nature* magazine.

Besides developing the finest ear for birdsong, Hudson was also fascinated and moved by the workings of adaptation and selection. His keen powers of observation led to numerous original discoveries, such as his note that the *chimango*, or common carrion hawk, "removes its young when the nest has been discovered—a rare habit with birds." His many references to birds' playfulness, his joyous descriptions of their prenuptial dances, and his fascinating observations of birds' motivation to migrate, all led to Hudson's explicit belief that birds experience emotions, appreciate beauty in their love songs and mating performances, and possess a soul. In this Hudson went beyond St. Francis—though he never tried to preach to the birds. The accumulated notes of this solitary rambler provide a glimpse of the great things Hudson might have accomplished as an American naturalist had he remained in the land of his birth.

Instead his expert knowledge led to introductions and offers from the Smithsonian Institution and the London Zoological Society. He collected hundreds of "skins" for them, traveling as far as Patagonia, Brazil, and Uruguay. Hudson was a good shot and table hunter in his day, with an attitude toward killing more like the American Indian than the Victorian collector. Later, however, a central tenet of his belief came to be that sparing bird life was not just good for the individual animal but for the soul of man; he angrily withdrew from the British Ornithologists' Union when it refused to stop egg collectors from decimating the nests of threatened species.

By 1869 Hudson was publishing a series of letters on the birds and other animals of La Plata in the London Zoological Society's proceedings, which enticed him toward a literary career in England. In 1874, his journals bulging from decades of field notes as an itinerant naturalist, Hudson left Argentina forever for London. Fifteen years as a pathetic failure and obscure literary hack followed, during which he was reduced to marrying a boardinghouse landlady much older than himself. *Argentine Ornithology* was finally published in 1889; it was Hudson's first bird book. Even then his collaborator, Philip Sclater of the London Zoological Society, received top billing, though Hudson contributed all the raw material and firsthand accounts of the birds—everything, in fact, except the classification and taxonomy.

Financial security and literary fame were finally won with the publication in 1904 of his commercially best-selling novel *Green Mansions*. Yet during all these years, marvelously written, passionately felt books and essays on natural history flowed prolifically from his pen. As one critic said of him, Hudson was a writer who said everything he felt, and felt everything he said. He transformed his deep-seated animistic beliefs about nature into a literature of powerful and prophetic dimensions. "For when man begins to cultivate the soil," he wrote, "to introduce domestic cattle, and to slay a larger number of wild animals than he requires for food . . . from that moment does he place himself in antagonism with nature, and has thereafter to suffer countless persecutions at her hands."

Yet few were as happy as the white settler Hudson had known on the Argentine pampas. For his rough existence was gilded with something better than any hope of future prosperity. From the moment of his induction he was engaged in a conflict, which Hudson saw as an incomparable inspiration to lead a healthy and interesting life. To the frontier feeling of challenge was added "the charm of novelty caused by that endless procession of surprises which nature prepares for the pioneer." Hudson called nature the feminine warrior, "a rill of pure water . . . where, for many years to come, [the pioneer] will refresh himself every day, and learn to feel (if not to think) that it is the sweetest rill in existence."

The stolid, sullen, immutable, always obedient nature that Hudson knew in England was changed on the pampas into a flighty, capricious thing, "a wayward Undine, delighting you with her originality, and most lovable when she teases most; a being of extremes, always either in laughter or tears, a tyrant and a slave alternately; today shattering to pieces the work of yesterday; now cheerfully doing more than is required of her; anon the frantic vixen that buries her malignant teeth into the hand that strikes or caresses her."

Frenzied as if by the indignities man subjects her to, she arrays herself in terrible black, and then, as if darkness were not terrifying enough, "she kindles up the mighty chaos she has created into a blaze of intolerable light, while the solid world is shaken to its foundations with her wrathful thunders." But storms are only the beginning. Nature calls her troops to action and sends a thousand torments to buzz in man's ears, sting his skin, suck his blood, poison his water, eat away his guts, and consume or kill his crops. She even denies him rest, with dreams of her monsters.

Yet the settler will not be beaten. The very charm of nature's challenges mitigates the despair. He slays her creatures. He dries up her marshes; he hacks down her forests and burns her prairies. He tames her rivers and turns her plains into orderly fields of corn and orchards of fruit-bearing trees.

Nature sends in reinforcements—armies of creeping

things and clouds of flying ones. Mice, crickets, moths, locusts. They destroy man's granaries. They devour his crops and trees, and turn his fields into barren, desolate things, cracked and scorched by the pitiless sun. And when the dead are fallen on both sides, nature "dries her tears and laughs again; she has found out a new weapon: out of many little humble plants she fashions the mighty noxious weeds."

As if by a miracle, Hudson said, the weeds take over everything. If man cuts them down in the morning, they have grown back by nighttime. "With her beloved weeds she will wear out his spirit and break his heart; she will sit still at a distance and laugh while he grows weary of the hopeless struggle; and at last, where he is ready to faint, she will go forth once more and blow her trumpet on the hills and call her innumerable children to come and fall on and destroy him utterly."

Yet to W. H. Hudson, a man who finished his life in a fall from his horse or met death trying to ford a swollen stream was better off by far than one who died of apoplexy in a Victorian dining room, or fell over dead in a London countinghouse.

There matters stood until 1920, two years before Hudson's own death. The time of gilded optimism in the Machine Age was past. Instead of human progress and the bloom of civilization, the century ushered in total warfare and the severing of man's vital links to the natural world. The open and free pampas were *Far Away and Long Ago*, as Hudson so aptly called his autobiographical memoir. In his last years, now Britain's grand old champion of bird conservation, Hudson decided to reissue his youthful ornithological field notes under the title *Birds of La Plata.*

He cut all the Sclater material, claiming the book's original taxonomy had been "out of date as soon as published" because new species had moved into the area and others had left or were exterminated. The numerous references in *Birds of La Plata* to species cruelly and stupidly hunted to extinction lead one to believe Hudson added considerable material from his later perspective. He was probably correct by that time in his bittersweet assessment that the book's chief literary interest was that W. H. Hudson had written it. His new introduction, however, noted the ornithological reason for republication of "these little bird biographies": no other books on the same subject had been written in the intervening decades. Certainly no better book on the ornithology of "the great bird continent," as Hudson liked to call South America, has appeared since. To this day Hudson's methodology in the field is the model for any wildlife biologist or serious bird-watcher. His strong, clarion voice is the envy of any writer of natural history. And with the transformation of the pampas, many of the birds were gone for good.

Hudson lived long enough to deal with his own work in retrospect. He could see the prophetic value of his life histories. Along with *Far Away and Long Ago*, they completed his journey back home. Life, the living bird, was the thing, he said, "not the dead stuffed specimen in the cabinet." For once the life goes out of bird, or man, for that matter, "what is left is nothing but dust." Hudson felt "a pang" as he wrote these words over the grave of "that land so rich in bird life, those fresher woods and newer pastures, where I might have done so much." If humans are to make peace with nature, on which human survival ultimately depends, we can find no better guide than W. H. Hudson, who loved life on earth so profoundly.

9

Exploring a Tropical Reef with William Beebe

NO EXPLORATION of the American tropics would be complete without a dive into its lush turquoise seas. The vibrant abundance of tropical life forms does not stop at the shoreline but quickens its pace among the multistoried coral reefs. Reefs are often called marine jungles, and this is apt in the sense that they rival the rain forests for richness of species and clearly outdo them in their amazing colors and the intricacy of their physical structure. The reef is a crowning architectural achievement of tropical evolution and a facet of tropical nature where facts and beauty are often impossible to separate.

The living coral that produces the lime material of the

reef is neither a plant nor a fish but a colonial coelenterate of the muscular Anthozoa clan (phylum Cnidaria). Which means it is a kind of polyp that spends it life rooted to one spot, secreting a stony skeleton that provides housing, hunting habitat, and protective shelter for itself and for uncounted kinds of algae, snails, worms, mollusks, and fish. Although corals come in a staggering variety of shapes and sizes, the basic plan is like a balloon inflated inside a hard casing. The individual coral polyp has an internal body cavity opening to the outside only by its mouth, which is surrounded by tentacles armed with specialized cells for stinging, tangling, or gluing food—mainly tiny zooplankton.

Coral polyps enter into a pact of mutualism with certain single-celled algae, known as zooxanthellae. These tiny plant filaments live inside the coral polyps, which are transparent to allow in the sunlight the algae need for photosynthesis. The zooxanthellae give corals their fabulous palette of colors. More important, the corals take oxygen and organic carbons (sugars) cast off as the waste products of the algae's photosynthesis. Under ideal conditions—clear water, sufficient sunlight, optimum temperatures—the host corals gain all the energy they need from the tenant algae living within them. They use this energy to manufacture and deposit limestone at a faster rate than erosion destroys it. Thus the symbiotic relationship between coral polyps and algae is key to all coral reef formations, and explains why coral reefs thrive in an environment best described as an underwater desert—areas without suspended nutrients and sediments, which would reduce the rate of photosynthesis and coral formation.

To this stable, rooted, living stone structure are attracted all kinds of plants and animals for shelter, living space, and food. Coralline algae, for example, are red seaweeds that wedge between real corals and encrust them-

William Beebe's newfangled diving gear enabled him to explore at close hand the coral reefs of the tropical seas.

selves with lime, soon forming a structural component of the reef system comparable to a cement that bonds loose fragments and dead corals. Algal turf grows over coral like an underwater lawn, made up of bright green filamentous algae, the staple diet of many kinds of snails, sea urchins, and fishes. The variety of invertebrates and bony and cartilaginous fishes supported by a reef complex is truly spectacular—an estimated count of reef species is 950,000, or one of every four marine creatures. A coral reef is not only a colossal community of corals and algae but an underwater city of seafood, as complex as New York but in brilliant technicolor.

The historical difference between jungle and reef in terms of scientific exploration is the difficulty of an oxygen-breathing human staying underwater. Except in the case of very shallow reefs, diving with one chestful of air cannot provide meaningful data. Breakthroughs in diving technology came in the 1920s and belonged to an American scientist, William Beebe, who became the first human to explore the underwater environment.

At first Beebe submerged in a diving helmet, ancient ancestor of today's scuba-diving gear. This contraption was developed for commercial salvage divers. It consisted of a heavy, conical copper helmet fitted to rest on the shoulders, with a hose connection on the right side and two oblique glass windows in front. Around the edge was a flange on which four flattened lead weights of ten pounds each were hung, to keep the helmet from floating away. An ordinary garden hose led back to the surface, where an associate in the bow of a schooner or launch had to work the long iron lever of an automobile tire pump to send air down. This succeeded remarkably well and safely down to a depth of about forty feet as long as there were no kinks in the hose. "The sensation just above water," Beebe wrote in *Arcturus Adventure*, "is of unbearable weight, but the instant I immerse this goes and the weight of the helmet with all the lead is only a gentle pressure, sufficient to give perfect stability." Beebe downplayed the physical risks of diving with hose and helmet. If "serious danger threatens or the pump should go wrong for any reason, I have only to lift up the helmet, duck out from under it and swim to the surface."

To enter into this new realm required no practice, special skills, or elaborate preparation. To the contrary, Beebe considered the only thing a diver absolutely had to have was a healthy capacity for wonder. "If one dives and returns to the surface inarticulate with amazement and with a deep realization of the marvel of what he has seen and where he has been, then he deserves to go again and again. If he is unmoved or disappointed, then there remains for him on earth only a longer or shorter period of waiting for death; there can be little worth while left in life for him."

William Beebe was a Brooklyn-born Columbia graduate who became director of tropical research for the New York Zoological Society, a post he held for thirty years until his death in 1952. An ornithologist by training and a wanderer by temperament, Beebe branched into oceanography in the 1920s, at a time when modern diesel-powered research vessels equipped with trawls (for sampling marine life) and dredges (for sampling the bottom) made the seas an exciting new scientific frontier.

With his diving gear and new enthusiasm, Beebe was soon exploring the ocean depths. He made expeditions to the Galápagos Islands in the Pacific, to the Sargasso Sea in the mid-Atlantic, followed Humboldt's Current, and did a major assay of sea life off Bermuda. Once, he followed a major current rip in the Pacific for several days. A rip or drift line is a place in the ocean where two flows of water come together, "the meeting place of great ocean currents," forming a giant liquid wall. Beebe was first to follow and closely observe such a formation and to realize that as the mini-wanderers known as plankton drift into the rips, immense numbers of marine creatures congregate there to feed: "The thread-like artery of the currents' juncture seethed with organisms—literally billions of living creatures, clinging to its erratic angles as though magnetized. The floating, drifting world of ocean life was, of course, irresistibly swept there." With so much "food, manna, ambrosia, in stupendous quantity, to be had for the taking," the swimming hordes of fish, reptiles, and marine mammals arrive to take advantage. Trees and shrubs swept into the sea from land and picked up by currents end up in drift lines, so they are also major vectors of flora distribution. Today oceanographers are learning that such rips and drift lines also play a major role in the global transport of marine pollution.

In addition to a rare knack for bringing nature to life in his writings, Beebe's idea of science extended well beyond the conventional scientific portfolio of solving problems and understanding phenomena. The seasoned investigator, he believed, should view the organic world as "an inexhaustible source of spiritual and esthetic delight." Especially in colleges, Beebe thought, "We shouldn't allow biology to become a colorless, aridly scientific discipline, devoid of contact with the humanities." It saddened him to think that when death knocked on his door, "I may find myself among so many professional biologists, condemned to keep on trying to solve problems."

In studying the oceans, Beebe found not only a portion of the earth hitherto impenetrable, nor merely a new and unlimited source of scientific data, but a poetic return "to an older home, comparable in no way to aerial penetration and infinitely more remote and fundamental than our air-breathing life today upon dry land." The very fact that humans and other living things are mostly composed of water should serve to remind us of the origins of life in the pre-Cambrian seas, long before the first forest ever mantled a river valley, before the first prairie covered a plain, before the first animal ever equipped with lungs took its first gulp of air. "This seems to me a very wonderful thing, to walk about on land today, vitalized by a bit of the ancient seas swirling through our body," Beebe wrote. "It is somehow of a piece with the stars and time and space—something to be very quiet and thoughtful about, and proud of."

In the marine kingdom, Beebe said, "most of the plants are animals, the fish are friends, colors are unearthly in shift and delicacy; here miracles become marvels, and marvels recurring wonders." But the most wonderful marvel of all was that the oceans, our ancestral home, gave the earth "an amazing similarity to a living organism." In the same way that blood courses through our arteries and veins, bringing fresh energy and carrying off wastes, the oceans are the earth's circulatory system, cleansing our planet while refreshing land and life. Beebe's idea that the earth itself is akin to some great organic being would be taken one step further by René Dubos, Jacques Cousteau, and others, proposing an actual living earth as a spiritual basis for environmentalism.

In 1930 William Beebe became the first person to reach the deep ocean floor. He did so in a small steel diving chamber he dubbed the bathysphere, which he had designed with his engineer-partner, Otis Barton. The hollow steel sphere measured only four feet nine inches in diameter, with walls one and a half inches thick and a circular entry hatch fourteen inches across. The inside was just large enough to contain Beebe and Barton, oxygen bottles, chemicals to absorb moisture and carbon dioxide, a telephone, still and motion picture cameras, and a spotlight to illuminate the depths. Weighing in at about five thousand pounds, the bathysphere was lowered from a winch-equipped barge into the sea at the end of a half-mile-long steel cable with a twenty-nine-ton break strength. On June 6, 1930, Beebe and Barton crawled head-first into the bathysphere, listened as the four-hundred-pound door was bolted down with a sledgehammer, waved goodbye through the eight-inch windows, and were hoisted into the sea off Bermuda.

The physical properties undersea included "endless darkness, perpetual cold, and everlasting and terrific pressure." At 300 feet a leak developed under the bathysphere's hatch. Beebe knew that as they went deeper the pressure would increase against the hatch, probably stemming the leak. He ordered the dive to continue. The leak did in fact stop. But at 800 feet Beebe had a sudden, inexplicable feeling that something was about to go wrong, and ordered the bathysphere hauled up. Over the next four years the two men made numerous successful dives, with nothing more critical happening than once losing telephone contact. In 1934 they reached a depth of 3,028 feet, well into the "perpetual night" of the deep.

Harry Houdini could not have attracted more attention. Beebe's submarine adventures caught the imagination of the American public, and his breezy personal account of reaching the ocean floor in *Half a Mile Down* stirred popular enthusiasm for ocean research. From this moment on, William Beebe was that rare modern American popular hero, the scientist-explorer. His bathysphere is still on display at the New York Aquarium on Coney Island.

Beebe wrote more than a dozen popular natural history books about his expeditions, which took him to all corners of the intertropical zones, on land and sea. His other books include *Jungle Days*, *The Edge of the Jungle*, *The Arcturus Adventure* (about his expedition to the Galápagos Islands), and *Two Bird-Lovers in Mexico*. Beebe wrote with a delightfully upbeat Jazz Age swing, never fearing to improvise when he had to. Here he is, for example, describing the inner architecture of living sponges: "Within my field of view were two oblong caves, etched deeply into the hills, and from these perforated expanses swept downward into the awful gulfs of out-of-focusness."

In *Beneath Tropic Seas* Beebe recounted the tenth New York Zoological Society expedition, which he led to Port-au-Prince, Haiti, in the winter of 1927. The chief object of the expedition was "to study at close range and at first hand by means of a diving helmet the life of a coral reef." In his diving helmet Beebe could stay down as long as his assistant on deck pumped air from a large tank. Beebe was also at that time developing the first waterproof housings for still and motion picture cameras, in order to record the reef life, albeit in inadequate black and white. In addition, he brought two heavy-duty submarine electric lights to light up the water at night. In four months Beebe took some "300 strolls" along the Lamentin Reef during the Haitian expedition.

Beebe wastes no time in getting his readers wet on page one: "You are standing on a metal ladder in water up to your neck. Something round and heavy is slipped gently over your head, and a metal helmet rests upon your shoulders. . . . You wave goodbye to your grinning friend at the pump, and slowly descend, climbing down step by step. For a brief space of time the palms and the beach show intermittently through waves which are now breaking over your very face. Then the world changes. There is no more harsh sunlight, but delicate blue-greens with a fluttering of shadows everywhere. Huge pink and orange growths rise on all sides—you know they are living corals."

In his diving helmet Beebe could remain submerged for several hours, observing the fascinating reef life close at hand. Usually after forty minutes, however, a "gentle tug" on the hose would come from the deck of the research schooner *Lieutenant*, where "another impatient adventurer" was waiting to take his place. Beebe would regretfully turn back to the surface, "troubled by a sense of loss" that so many years of his life had passed without his "knowing of the ease of entry into this new world."

Before Beebe and his breathing apparatus, the world of the underwater coral reef was not at all easy to enter. In his *Beagle* days, Charles Darwin had seen "enormous areas in the Pacific and Indian Oceans, in which every single island is of coral formation." But he studied them, it seems, mostly from the deck of the ship. In *The Structure and Distribution of Coral Reefs*, Darwin did not attempt to recount the life of the reef but rather classified reefs geologically and explained his views on their formation.

Darwin found three kinds, or "classes," of coral reefs. The first was the fringing reef, where the corals followed the contours of the shoreline. Fringing reefs are limited by their location in several significant ways. Temperatures, for one thing, are higher in shallow water; corals are temperature-sensitive creatures. Their rate of calcification, or limestone deposition, drops above 84 degrees Fahrenheit. Tide levels are also a greater factor near the shore, as is wave action. Erosion can thus have an important effect on the fringing reef. Close to shore, deposition and sedimentation from rivers can also affect fringing reefs, both in terms of water turbidity and the addition of nutrients. Corals love their water neither murky nor too rich. Whatever the reasons, fringing reefs often are characterized by a lower ratio of stony corals to other sedentary species more tolerant to proximity to terra firma, such as sponges, algae, and the so-called soft corals.

Coral formations grow outward, away from the shore, so that fringing reefs gradually become barrier reefs, Darwin's second class. The barrier reef lies some distance offshore, parallel to the coast. The lakelike water between the barrier reef and the shore is the lagoon, a word that somehow feels naked without the adjectives beautiful and blue before it. Lagoons are never very deep, and their widths vary enormously, from swimming-pool size to thirty miles or more across. A great number of reef creatures feed in the lagoon, then use the reef for habitation and protection.

Barrier reefs typically have a massive limestone base over a stable platform of volcanic rock. Geological borings have shown that the limestone may sink thousands of feet. Given the relatively slow rate of coral growth, this means that coral reefs have been around for hundreds of millions of years. Barrier reefs can also attain tremendous lengths; the Great Barrier Reef off the coast of Australia is 1,260 miles long and is said to be earth's only underwater formation visible from space.

Darwin noted that almost every Pacific Ocean voyager "has expressed his unbounded astonishment at the lagoon-islands," or, as Darwin preferred to call them by their Indian name, atolls. "Truly wonderful structures," Darwin called his third reef class, these circles, ovals, or rounded crests of corals, usually set in deep, clear waters, seemingly independent of any landmass. What causes atolls to spring up so far from the shores? Darwin wondered.

Darwin rejected the leading theory of the time, which was that atolls form atop submarine volcanoes. How could they, he said, if barrier reefs, the most common form, continue for tens and hundreds of miles? Darwin held instead that atoll reefs developed as a result of islands once fringed by coral reef gradually sinking into the sea. According to Darwin's conception, the island once surrounded by fringing reef subsided very slowly, while "the little architects" continued to build their great walls of limestone.

Darwin's theory of reef formation by slow subsidence has held up well to modern scrutiny. Acceptance of the theory of plate tectonics, for example, makes it much more likely that the ocean floor can sink. But Darwin's certainty that "we see in each barrier-reef a proof that the land has there subsided, and in each atoll a monument over an island now lost" is a bit overboard. While most coral reefs have indeed formed as a result of subsidence, reefs also exist in areas of long-standing stability and uplift. Some reefs unquestionably form around submarine volcanoes. In sum, coral reefs probably form in different ways in different parts of the oceans.

William Beebe dispensed with abstract theorizing in favor of a close-up angle on the reef: "You lean against a fretwork of purest marble while at your elbow is a rounded table of lapis lazuli on which are blossoming three flowers—flowers unearthly and which lean toward you of their own free will. Their petals are resplendent in hues of gold and malachite, and are fluted and fringed like some rare and unknown orchid. You reach forward to pluck one, and, faster than the eye can follow, the blossoms disappear beneath the fur of lapis velvet from which they seemed to sprout."

The reef corals themselves ranged from tiny one-inch pink trees and sushi-size discs to "massive brain [corals] as big as automobiles, and elkhorn [coral] forests twelve and fifteen foot high." Lamentin, or Sea-Cow Reef, as it is also known, is a barrier type of reef, lying parallel to the land and several miles offshore, with a deep channel in between. Beebe reported a "beautifully graded transition" from the land to the deep water, with coconut palms giving way to a fringe of mangroves "with their toes wet by the high tides." Next came sandy beach reaching beyond the tide line, followed by a wide lagoon zone thick with thalassia, or eelgrass. It was most likely the eelgrass that attracted the manatees, or sea cows, and gave the reef its nickname.

Rather abruptly this underwater pasture merged into the level and shallow inner side of the reef. There Beebe found creatures that did not occur farther out in deeper waters. "Under every bit of coral swarmed starfish—black, red, pied, grey, orange and purple. Sea-urchins vied with them in numbers—long, needle-spined chaps, short stubby club-spined ones, and others fashioned like chestnut burrs. With these were hosts of hermit crabs, small worms, infant fish and octopi."

Next came an abrupt drop to several fathoms—the elkhorn wall, so to speak, made of tall growths of the now familiar elkhorn and other branch corals. These grew "in a ghostly tangle of cylindrical, white thickets . . . quite impenetrable," wrote Beebe. "I ventured more than once to creep down into these tangles of coral branches, testing each before I put my weight on it." Beebe had to clamber down this coral maze with extreme care, not only for fear of getting slashed and cut about the body (which happened anyway whenever the sea was the least bit turbulent); he was more concerned that his air hose would become entangled and jammed in the coral. He explained, "In the open reef, no matter what happened, one could always lift off the helmet and swim up, but here there was a cruel, interlaced, cobweb of sharp-edged ivory overhead, and escape was possible only by slow deliberate choice of passage."

The prize proved equal to the risk. As he slowly wended his way down through stinging millipores to the ground corals, Beebe used his geologist's pick to chip off chunks of the outer layers. And there, "like jewels in a geode," he found "tiny trees, an inch or two in height," of the exquisite and rare pink coral. Beebe added, "I do not remember anything in my undersea experience which gave me more sheer aesthetic joy than spying out these beautiful bits of color—looking like the diminutive wind-blown pines of Fujiyama."

When the sea was too rough or the sky too cloudy for photography, Beebe broke off huge coral branches from the reef floor and sent them to the surface for analysis on the boat. He assumed that the bottom-dwelling corals, somber-colored in browns and greens and dull purples, did not receive enough light, nutrition, or protection, to develop luxuriantly. But he soon found otherwise.

When the scientists of the Tenth Expedition began to break open the coral debris sent up from the reef floor, they found "Aladdin's caves everywhere, and our eyes were flooded with imprisoned rainbows and spectrums." From inside the corals blossomed violet flower worms, sunset pink shells, and crabs "hiding in filched shells, which in turn were in coral chambers from which there was no escape, and the colors of their legs and eyes defied human names."

Where the corals thinned out, softer corals grew in the shapes of nets, fans, ferns, and plumes. Here Beebe closed in on the gorgonia corals "covered with a dense polyp fur of clove brown." The polyps not only looked like part of one big furry creature, they also acted in unison, like one being. Beebe actually saw them share an emotion: "Some of the fluffiest stung at a touch, and, at the same touch, withdrew every polyp, leaving bare the rich purple trunk and branches—an emotional autumn of fear which swept over the full-blown foliage swiftly as a shadow."

The tropical reef is the veritable kingdom of fishes, and Beebe's principle scientific reason for the 1927 expedition was to construct a list of fishes in and around the reef. Anyone who has dived near a tropical reef, or collects tropical fish in an aquarium, for that matter, knows that it is impossible to describe in words the magnificent array of colorful fish Beebe had before his eyes. Each kind with its own color pattern, colorful personality, and niche in a seemingly endless food chain. For the reef itself, as we have noted, is destitute of nutrients, and everything that lives there lives according to the iron law: big fish eat little fish, little fish eat salad.

Beebe watched, enchanted and barely breathing. "Before long . . . a mob of huge parrotfish came into full view, working slowly toward me, feeding and idly wandering about as they came. They drifted around a coral spur, but before the last straggler vanished, the vanguard appeared again out of the distant brilliance, and now their numbers were augmented. I counted up to one hundred and thirty nine, and then realized that three hundred would be within reason. None were less than a foot, while most were more than two feet in length, and at least twenty measured a full yard."

The parrotfish wrenched off the heads of corals looking for food, and through the limestone debris Beebe saw tiny demoiselle fish swimming out from their coral homes to attack the parrotfish, in true David-versus-Goliath fashion: "To see a three-inch black and yellow fury driving full force against the side of these blue enameled giants was to see courage at its height."

Beneath Tropic Seas is filled with fantastic fish stories—all true. Using his submarine night light, for example, Beebe captured a rather large jellyfish. When he got it into the laboratory light, he could plainly see that fish were living inside the jelly's poison tentacles! "How such a habit could ever have been inaugurated and established is inconceivable," Beebe commented. Yet it proved to be a routine phenomenon around his sunken lights. A photo in *Beneath Tropic Seas* shows a giant jellyfish with its tentacles cut off and 355 fish living inside. Beebe was able to observe the jelly dwellers darting in and out of their exquisitely dangerous apartment: "Twisting and wriggling about in the wake of the jellyfish were four slender, pink tentacles, ready at a touch to shoot one's hand full of nettles. It was exciting to see the fish manoeuvre for a few seconds before they dashed in to safety between the tentacles. It was like running into an open door before which was suspended a cluster of swaying, twisting live wires." Sometimes the timing of the fish was off, and the "nettle lariats" instantly turned them from jellyfish tenants to jellyfish food.

Beebe listed the reef fishes according to an idiosyncratic "occupational" classification system he had previously devised during his Galápagos expedition, grouping fishes by the way they work and live—which makes a good deal more sense to the average reader than Latin nomenclature. It also emphasizes the connectivity of the tropical reef ecosystem. Beebe's list is reproduced here not only to demonstrate some of those ecological niches but also to illustrate the awesome bounty the future marine biologist working in the Caribbean can anticipate:

FREE NOMADS
Sharks
Eagle Rays
Carangids
Tarpons
Cornetfish
Mackerels
Groupers
Gars
Barracudas
Puffers
Dolphinfish

SQUATTERS
Godies
Blennies
Morays
Eques

BALLOONISTS
Young Bumpers

VILLAGERS
Demoiselles
Butterflyfish
Grammas

AERONAUTS
Sargasso-fish
Young Triggers
Pipefish
Seahorses
Abudefdufs
Tripletails

SPONGE PEOPLE
Amias
Blue-lined Blennies
Garmannia

SAND CRAWLERS
Skates
Flounders
Batfish

GRAZERS
Parrotfish
Triggers
Surgeons
Angelfish

PERCOLATORS
Snappers
Wrasses

FLYERS
Flying Fish
Halfbeaks
Gurnards

SURFACE MOBS
Anchovia
Silversides

Beebe's research, it should be noted, occurred long before contemporary sensitivity to scientific methods. He used every available means to collect 270 species from 6,122 specimens, including hook-and-line fishing, netting, seines, set-traps, air rifles, harpoons, poisons, dynamite, and even high-explosive bombs dropped from airplanes. Beebe's scientific heirs recognize that such blunderbuss ways damage the very attitudes of wonder and respect for life he tried to foster. Moreover, researchers today warn of a precipitous decline in coral reefs worldwide that absolutely precludes such harmful human disturbance, especially in the name of knowledge.

Coral reefs, indeed, may be dying off even faster than tropical forests. One gloomy forecast by U.S. National Oceanographic and Atmospheric Administration scientists predicts the destruction of all coral reefs within thirty years. Perhaps such predictions are exaggerated by inexact computer models of the future global climate. Even so, it is beyond doubt that coral reefs—structures that handily survived the two great biological dieoffs in earth history—are being mortally damaged today at an alarming rate.

Who is killing the world's great reefs? The factors are global as well as local. On a planetary level, the common greenhouse effects of increasing temperatures and rising sea levels bode ill for the corals, highly sensitive to both. There may also be a downward spiral effect. The carbon dioxide that humanity is packing into the late-twentieth-century atmosphere dissolves in seawater and becomes carbonic acid. The limestone casings of corals can absorb huge amounts of the acid, so flourishing reefs act as geochemical buffers, reducing the effects of global warming. If the system is overwhelmed, however, even more coral reefs will die sooner.

There is also increased danger of flooding lagoons with rising sea levels and increased storm surges. Lagoons, which form between land and reefs, also play a regulating role by acting as evaporation basins, leaving behind salt crystals. This exchange helps the seas maintain their proper salinity in the face of all the minerals washing into the oceans as sediment carried by the world's rivers. And this is where the crisis of the tropical forests and the reef crisis meet.

Everything we do to the land affects the seas. When tropical forests are clear-cut or burned, the land loses its holding power and the soils erode. Sediment is washed into the rivers, then the seas. Formerly clear coastal waters become turbid; the corals cannot survive the reduction in water clarity, and suffocate. Thus tropical deforestation ultimately damages coral reefs.

Finally, human numbers and human technology seem set against Beebe's coral cities beneath tropic seas. With more commercial shipping, fishing, and recreational boating, damage to corals from shipping impacts is multiplying. The growth of international tourism has unfortunately loosed huge schools of first-time divers on the Pacific coral reefs, who break off corals for souvenirs, and hordes of uneducated fishermen, who drop their anchors onto the brittle corals millions of times a year. The runoff from untreated sewage in tropical countries with burgeoning urban populations feeds the growth of coral-choking algae blooms.

As we have learned more about corals reefs since the time of William Beebe's adventures, we have become aware of the acute fragility of the coral reef's environmental zone. Flourishing in underwater deserts, tough as nails down through geologic time, coral reefs are staggeringly productive biologically but highly vulnerable to changes in temperature, sunlight, and salinity.

Early in his career, beginning a round-the-world trip to study wild pheasants, William Beebe was given "one of the most valuable hints I have ever had." Later, when he was the first human to study coral reefs close at hand, he thought of it often. In most peak experiences in life, "it is not, as so many people think, the first few minutes which are the most wonderful. It is the subsequent gradual appreciation which develops," he wrote in *Arcturus Adventure*. It is easy to miss this protoconscious absorption of a journey or experience, and afterward "we long for just one moment of the actuality, so that this or that could be seen again and remembered more clearly."

The friend's advice was based on a sound realization of a well-known human weakness—familiarity breeds contempt. The recommended defense consisted of shutting his eyes "in the midst of a great moment, or close to some marvel of time and space, and convincing yourself that you are at home again with the experience over and past." What would you most wish to see or do if you could turn time on its head?

Underwater for the first time in his copper diving helmet, Beebe sat down on a convenient rock, shut his eyes, and recited his explorer's credo: "I am not at home, nor near any city or people; I am far out in the Pacific on a desert island sitting on the bottom of the ocean; I am deep down under the water in a place where no human being has ever been before; it is one of the greatest moments of my life; thousands of people would pay large sums, would forgo much for five minutes of this!"

Then he opened his eyes.

On the last night of the Haitian expedition to study Lamentin Reef, Beebe worked with his lights and watched from the gangway until his eyes hurt and his muscles cramped. He went to his tent to rest. But at midnight he returned to the boat and turned on the big two-thousand-candlepower light one last time.

We leave William Beebe, just as he left the tropical

reef, during a piscean flash of illumination: "There were the multitudes still milling about—searching, fleeing, fighting, mating. And I tried to imagine the centuries and millenniums that this had been going on before Haiti ever rose above the waters; and my mind went ahead to the ages when it would still be in full sway, eons after the last human being had completed his span of life, and I gave it all up. But I knew that as long as I could strive to take in the sweep and swing, and warp and woof of things, cosmically and universally, I need fear no touch of conceit for anything I might ever achieve."

10

Archie Carr: The Puzzlemaster

ARCHIE CARR was one of the founders of conservation biology, for which there is no Nobel Prize. Until his death in 1987, Carr was the world's leading researcher and authority on sea turtles. In his books, colored with a warm intelligence and lustrous, low-key wit, driven by his boundless curiosity, he chronicled an enviable life of indefatigable research and travels. He first roamed the tropical shores of the world's oceans as a naturalist, later concentrating his study on green turtles (*Chelonia mydas*) for nearly half a century at Tortuguero, which means Turtle Bogue in English—the last green turtle nesting beach of the western Caribbean, a sunny stretch of coconut-fringed sands on the unspoiled coast of Costa Rica.

Carr's study of sea turtles began after World War II. Heading into a time of cold war and organization men, Archie Carr was anything but a cookie-cutter scientist in a white lab coat. While others turned progress into our most important product, Carr roamed the coasts of tropical Africa looking for turtles, talking to fishermen, absorbing landscapes and wildlife. He persuaded the U.S. Navy to sponsor his research on turtles, with the intriguing notion of unlocking the secrets of long-distance animal marine navigation. Carr eventually convinced the country of Costa Rica not only to declare his Tortuguero study site a national park but to make it one of the first wildlife refuges to enforce laws against turtle slaughtering and to shoot encroaching dogs. Tortuguero was also one of the first places to make ecotourism work for conservation, as thousands made the pilgrimage there to observe and study the turtles Carr celebrated in his several books.

Archie Carr was a talented writer whose most famous work, *So Excellent a Fishe: A Natural History of Sea Turtles*, became an instant classic of natural history. He was not only a landscape painter in words and a spinner of yarns, though he was good at both. Carr was a first-rate scientific explorer who could communicate the excitement and thrills, no less the logic and methods, of his explorations. "When I was growing into my leaning toward the tropics," he wrote in *The Windward Road*, "it was the custom among sound scientists to inveigh against the having of what they spoke of as 'adventures' in the field. Adventures on an expedition were the sure mark of incompetence, they said, or of chicanery. . . . The thing is . . . adventure is just a state of mind, and a very pleasant one, and no harm to anybody, and a great asset if you use it right."

Archie Carr's studies of sea turtles in the Caribbean revolutionized conservation biology and eventually saved several species that seemed headed for extinction. (*University of Florida, Marshall Prine*)

Archie Carr used his adventures well. Each of his trips to the Caribbean was another installment in the piecing together of a large puzzle and in his lifelong efforts to solve, with little more than his wits and his hands, the evolutionary paths of the creatures he pursued—"the riddle," the "mystery," as Carr often called it, of life as a sea turtle. Carr's books are marvelous examples of clear scientific thought. His unflagging spirit of youthful adventure advertises a virtue the best naturalists possess, that of taking on nature as a kind of large outdoor game, which anyone with imagination, tenacity, and physical stamina can play.

So Excellent a Fishe is an account of one man's knowledge, observations, hypotheses, ingenious experiments and speculations on the evolution and survival of these 100-million-year-old ocean reptiles. Several species of them seem headed toward extinction in our lifetime; the rest are so reduced in numbers and interfered with by humans that, as Carr feared toward the end of his life, their life histories may never be unraveled. But a few of them may yet survive, thanks in no small part to Archie Carr's energetic conservation efforts.

Practically everything that was known about the life history of sea turtles when Carr started out in the early 1950s was a result of the female sea turtle's adherence to the ancestral saurian habit of laying her eggs in a hole in the ground. For a few weeks each year, she-turtles lumber from the foam, dig nests in the sand safely above the tide line, and lay their eggs at a depth unlikely to be desiccated by the heat of the tropical day or washed out by storms. Unlike the crocodilians, however, female sea turtles do not stay around to protect their young. Two months later the eggs hatch, the baby turtles crawl up to the beach level, and by some instinctive reaction head immediately, though not necessarily directly, toward the water. If they make it, the rest of their lives, which some think could be two hundred years, is spent in the sea, except as they may become stranded by storms or grow up as females to repeat the nesting cycle.

Thus although sea turtles spend 99 percent of their lives in the sea, the pelagic life history of sea turtles has been virtually unknown. Where do they roam? The world's oceans cover two-thirds of the planet's surface—how do turtles orient themselves in all that water? How do they navigate? Do they follow certain migratory paths, like sooty terns or salmon? Do newborn turtles find and use the same sea lanes as 150-year-old turtles? Do the females return to the beach of their birth to nest? If so, have they inherited an "unswerving attachment to ancestral grounds" or only what is known as imprinting—the stamp of earliest memory, aided by the look, feel, smell, and taste of the natal shore and the natural signposts leading to it? And so on. To think you know the biology of a marine creature like the sea turtle on the basis of a three-week terrestrial nesting period for the mature females is like a man thinking he knows women from having once seen a Mae West movie. It was these basic questions about sea turtles that Carr set out to answer.

"To learn anything about the natural history of a wandering seafarer such as Chelonia," Carr says in *So Excellent a Fishe*, "the main problem is to keep the animal in view." There are but two moments in the life of a sea turtle when the student can count on making contact: when it hatches, and when the female goes ashore to nest. "Everything else," observed Carr, "is done away off somewhere out of sight, and has to be reconstructed by deduction from fragments of observation."

With no better place to begin, Archie Carr started with the peculiar pilgrimage of an ocean reptile back to the land its ancestors left 100 million years ago.

"Everyone," Carr remarked, "should see a turtle nesting." The beaching of the females, or *arribada*, as it is sometimes called in Spanish, is one of the most impressive events in the entire biological world. Between April and June, during times of high inshore winds and rough seas, always in daylight, along one ninety-mile stretch of beach in Mexico, for example, more than forty thousand female Atlantic ridley turtles may stagger from the sea to dig a precisely measured, flask-shaped nest and lay roughly one hundred eggs. Although not all prone to mass emergences, the females of all five kinds of sea turtles (green, loggerhead, ridley, hawksbill, and leatherback) make their nests with the same "machinelike instinctive pattern," digging as deep as the turtle's back flipper is able to stretch. Sometimes during the Mexican *arribadas*, solid miles of ridleys, instinctively secured into their ovipositing duty like robots, make a hardshell gridlock.

Why do such great gangs form? What is the biological utility of concentrated nesting? Setting aside for the moment the idea of mating congregations as gene supermarketing, Carr turns to the most obvious answer: nature's great and ghastly game of predation and survival. Always in attendance at such turtle egg-laying fiestas are the egg-eating predators awaiting their groceries: coyotes and feral dogs, snakes, skunks, and all the bird scavengers that accompany them. They are traditionally joined by human egg collectors and the turtle turners, or *veladores*, who upend the females, rendering them helpless to escape butchering.

Two months after the nesting, an even more daunting gauntlet of predators gathers to pounce on the newborn hatchlings as they surface from the nest and take their first steps toward the sea. To Carr, the mayhem and slaughter accompanying the hatchling turtles' first dash to the water was an "ecological monstrosity." Those who have witnessed this scene *in vivo* can never forget it. Thousands of turtle hatchlings, each no larger than a twenty-five-cent piece and seemingly made of dull black rubber—perfect only in their vulnerability—are picked off as they race to the surf across a beach hungrily patrolled by dogs, hawks, possums, peccaries, monkeys, snakes, crabs, buzzards, and even jaguars and pumas. The beach, floodlit by tropical moonlight, is turned into a killing field, while just offshore in the silvery waters wait the host of hungry sea hunters: redfish, sea trout, snook, sharks, snappers, and mackerel. The morning light reveals the carnage, while sated predators lick their chops and scavengers mop up the massacre. The picture is not pretty, and can swiftly alter a naive association of the "natural" with the good. Grown sea turtles have few enemies outside of humans, Carr noted. But "the whole world seems against the hatchling."

Evolution appears to have offered sea turtles a deal somewhat worse than Faustian: survival through the eons in return for the mass sacrifice of their newborns. But it has worked, which is always the ultimate test in evolution. By concentrating the release of their offspring to a relatively short burst of time on single stretches of beach, sea turtles have apparently struck on a reproductive strategy that, if not outwitting predators, perhaps overwhelms them with numbers.

All five kinds of sea turtles lay more than once each season. Carr found that green turtles, for example, lay on the average four times. That makes each female responsible for about four hundred eggs. More her physiognomy probably could not carry. Fewer would probably not constitute the critical mass necessary (minus human predation) for sufficient survivors. A lot of natural selection must go into reproductive behavior when the parent does not provide child care. "The desperate plight of the young [turtles] on shore," Carr writes, "just adds a bit more proof to the proposition that the hundred eggs a turtle lays is a package pregnant with meaning and history."

We mammals, who lord it over the rest of creation with our milky ways and lengthy nurturing of our young, are wont to look down our appendages with a certain contempt for those orders of animals that cut and run when it comes to reproduction. Archie Carr's descriptions of turtle nesting and turtle birthing, processes "packed with ecology," restore a certain respect, even sympathy, for a creature across a great divide from us in adapting for survival.

One year at Tortuguero, Carr and his colleagues Larry Ogren and Harold Hirth set out to learn some of the subtler ways that eggs laid in a clutch produce more new mature turtles than eggs laid singly. By digging up to a nest from one side and replacing the sand wall with a pane of glass, they were able to observe exactly what takes place inside. "The first young that hatch do not start digging at once but lie still until some of their nestmates are free of the egg," they wrote. "Each new hatchling adds to the working space, because the spherical eggs and the spaces between them make a volume greater than that of the young and the crumpled shells."

In the windowed nest, Carr and his coworkers witnessed "a witless collaboration that is really a loose sort of division of labor." The turtles on top scratch down the protective ceiling of the nest, backfilled and pressed down by the mother before returning to the deep. Those around the sides dig outward, undermining the walls. The turtles on the bottom accomplish two roles. First, they trample and compact the sand falling from above. Second, "they serve as a sort of nervous system for the hatchling superorganism, stirring it out of recurrent spells of lassitude. Lying passively for a time under the weight of its fellows, one of them will suddenly burst into a spasm of squirming that triggers a new pandemic in the mass. Thus, by fits and starts, the ceiling falls, the floor rises, and the roomful of collaborating hatchlings is carried toward the surface." Soon the baby turtles erupt onto the surface and take the field like a swarming football team, though without a coach to tell them in what direction to run toward the goal.

The vast majority of hatchlings never catch their first wave. The plight of the turtle hatchlings in these very few minutes is desperate: "The moment of maximum peril in the whole life of the species," Carr called it. But if a few make it, the contribution of this mindless teamwork, or proto-cooperation, cannot be discounted. The turtle siblings appear to work together in emerging from the nest as a survival unit, though operating completely by instinct, without any suggestion of altruism. "Their petulance at being crowded, jostled, and trod upon make them flail about aimlessly. It is the aimless flailing that takes them steadily up to the surface of the ground."

To find out if such instinctive teamwork enhanced baby turtles' successful emergence, Carr buried twenty-two single eggs, then kept track of their fate. Only six of the lone hatchlings ever reached the surface. All were too fatigued by their efforts to reach the water. Carr thought there was no doubt that unconscious cooperation continued as the hatchlings raced for the water. For one thing, rests and hesitations were fewer because turtles from a group of one hundred kept bumping into one another. "If a sprinkling of hatchlings has stalled for a time, and a nestmate comes charging up from behind and touches one of them, the one touched springs into action, the action spreads, and the whole group scrambles away off toward the sea as if you had wound them up like toys and set them down together."

Carr associated the remarkable ability of the hatchlings to locate the sea with the female's ability to return to the sea after laying her eggs (she never follows the trail she has made incoming), and distinguished this "fairly fancy guidance process" from the orientation of adult turtles in long-range migratory travel. Carr suspected that both sea-finding and long-range navigation were probably "composite sensory processes," where a basic signal or main guidepost—such as light or compass sense—was supplemented by local signals. But in the trip from nest to water, "orientation seems to be basically a tendency to move toward a special kind of illumination or away from the lack of it." In other words, sea-finding involves vision and light but is not a simple tendency to move toward, say, the sun or moon.

This was established by Carr, his graduate students, and many others, in a number of experiments during the 1960s which used everything from blindfolds, hoods, and sunglasses to photographic flashbulbs and spotlights. In one of Carr's zanier attempts to solve the puzzle of sea-finding, he scooped up a boxful of hatchlings from the beach at Tortuguero, climbed into a waiting two-seater airplane, flew them across the width of Costa Rica, and released them a few hours later on a Pacific beach out of sight of the water. Were they disoriented by the sea change? Not at all. The newborn turtles still marched resolutely toward the sea, demonstrating that they apparently were not born with compass sense—i.e., the instinct to go in a particular compass direction.

The stubborn sea-finding ability reaches its final stage in the sudden and violent encounter of the hatchling with the extremely dynamic surf line. As soon as the baby turtle reaches wet sand—even before it reaches the actual waterline—it breaks into bursts of swimming strokes, like the arms of a clock spinning crazily in a Charlie Chaplin movie. Animal behaviorists call this a "releaser effect."

Carr studied it by releasing hatchlings near especially clear water, where he could take a boat alongside and observe their reactions. He found that the swimming instinct at first comes and goes while the hatchlings are alternately lifted and stranded by the advance and ebb of the sheet flow. As the turtles reach deeper water, they struggle forward a few feet under water, then emerge to breathe and look around, before going down and moving ahead again. None of this appears to be learned. Within a few minutes of leaving the nest, the baby turtles swim with the same authority, if not yet the full power, that will get them through the next century.

But as the young chelonians breach the last crash of surf and disappear, all knowledge of the newborn turtles ends. No one will see them again until they weigh about ten kilograms and have reached the approximate diameter of basketballs, probably as yearlings. Where the newborns go, what they eat, how they survive constituted the baffling "Lost Year" puzzle, as Carr dubbed it. Out of sight, though perhaps not out of mind, there was simply no way to keep the survivors of the year's hatch in view in the vastness of the world's seas.

Carr decided that the best way to take advantage of the brief contacts between turtle and turtle-student was to "follow" the females. This could be accomplished by tagging: drilling a small hole in the back shell of the female, and wiring a small metal tag to her before she reentered the ocean. The tag was inscribed with a number and with the offer, in English and Spanish, of a reward to the finder who returned it to the Department of Zoology of the University of Florida at Gainesville, Carr's academic headquarters.

At first Carr and his students used "an inscribed oval plate of monel metal, wired to the overhanging back edge of the shell with monel metal wire." They caught and tagged several hundred green turtles at Tortuguero, while in Florida they marked loggerheads, green turtles, and ridleys with the same tags. It soon became obvious, however, that most of the tags were becoming lost before the turtles even left the near-offshore nesting grounds. "It seemed impossible," Carr noted, "but time after time a turtle would return [to the nesting beach] tagless after an absence of less than two weeks. Such turtles could be recognized by the empty holes in the back edge of the shell," where the wire had been inserted.

Where did the females go between nesting forays to the beach? Carr didn't know. But one thing that occurred during this time, he noted, was "a lot of strenuous romance" just in front of the shore. It became apparent in the very first year of the tagging program that the loss of tags was the handiwork of the rutting males. "Sea turtles in love are appallingly industrious," Carr commented wryly. "The male obviously makes an awful nuisance of himself. Why the female puts up with such treatment is hard to understand." Carr then described the "three-point grappling rig" consisting of hooked claws on each flipper and a horn-tipped tail, with which the two-hundred-to-three-hundred-pound male turtle mounted the "smoothe, curved, wet, wave-tossed shell of the female."

Eventually he realized there was no way to prevent the males from demolishing the shell tags in these ferocious courtship embraces. From another researcher he heard that common-farm-cow ear tags, clipped to the back flipper to avoid interfering with the turtle's movements, stayed attached much better. The first shipment of these tags arrived at Tortuguero only a few days before the end of the first year of the tagging program. Carr rushed to the beach and worked constantly for the next four days, putting the new tags on forty females. Renestings showed little loss of tags. Four of the forty were returned from other countries—a good recovery percentage in a study involving long-distance travel. "We gave up the shell tags forever," Carr recalled.

The turtle tags did not say so, but the reward for return was five dollars, which Carr always "paid promptly and without any haggling." If a turtle tagged at Tortuguero was later hunted and killed off the Mosquito Coast, stranded on a Caribbean beach, or caught in a Florida Gulf Coast fishing trap, the return of the tag would feed Carr one more bite of information on the movements of sea turtles. And since five dollars was in those days a lot of money in the Caribbean—more, for instance, than the cash value of the turtle, most of which was then eaten locally—any tag found was likely eventually to find its way back to Gainesville. "I imagine," said Carr, "that the National Science Foundation [which supported the turtle tagging project] had misgivings over the reward item in the budget of my research project plan; but they sent the money, and I never spent so little to learn so much." Later on Carr would joke, "If my name goes down in the canons of zoology, it will be as the instigator of the five-dollar turtle-tag reward."

With the start of the tagging project in 1954, the simple tag became Archie Carr's most important investigative tool. It was not a high-tech operation. "It lacks the fascination of the gear of oceanographic research or of the apparatus of biochemical research," wrote Carr. "But it is an effective device all the same. Seldom can so much be learned from so little manipulation as a tagging project demands."

Four thousand two hundred mature green turtles had been tagged at Tortuguero by the time the first edition of *So Excellent a Fishe* was published in 1967. By 1983, when a revised edition appeared, some 20,000 turtles had been tagged. When Carr died in 1987 the number was already 35,000. And the turtle tagging continues today, perhaps the world's longest-running scientific study of an animal.

Several hundred of the first-edition turtles returned to Tortuguero for renesting; this showed that the colony nests at least three, and as many as five or even six, times per season, at an average interval of 12.5 days. In the second year of the tagging project, none of the she-turtles tagged the first year came to shore, but some arrived the second year, and more the third. This pattern continued, arguing for a two-to-three-year sexual cycle, perhaps changing as the turtle ages. In addition, Carr's tagging team marked off the beach in a fine grid and kept records of where turtles nested and renested, both within a single season and after the two-to-three-year cycle. In this way, evidence accumulated that green turtles are amazingly site-tenacious, often returning within an eighth of a mile of their previous nesting. Are these also the same sites where they were born?

The primary result of the tagging project has been for the first time to build a scientific basis for reconstructing the long-distance travels of sea turtles. The very first nonlocal recovery was from a turtle harpooned by a Miskito Indian from Puerto Cabezas, Nicaragua, two hundred miles from Tortuguero, and returned to Gainesville by the local parish priest. One hundred and seventy-five "five-dollar tag returns" followed from every part of the western Caribbean and as far away as the Guajira Peninsula off Colombia, the waters off eastern Cuba, and the Marquesas Islands near Florida.

Since the turtles were tagged at the end, so to speak, of their journeys to Tortuguero, Carr had to use reverse logic in deriving tentative conclusions. Here is an example of one such procedure. First, the preponderance of tag returns from the Mosquito Cays off Nicaragua, combined with the dearth of returns from Costa Rican waters in nonnesting seasons, demonstrated that there was likely no resident colony near Tortuguero. Second, the lack of a single return from any other landfall (combined with years of fruitlessly scouting the coastline for alternate nesting beaches) seemed to prove that Tortuguero was the only nesting beach for green turtles in the entire western Caribbean. Third, the evidence of turtles returning in two- or three-year cycles to the same beaches, as previously mentioned, was evidence of site-tenacity. Taken together, these three tentative conclusions led Carr to the insight that the turtles mating and nesting at Tortuguero must travel there from quite different and far-flung places. If so, the geographic isolation of widely separated green turtle populations "is not true genetic isolation at all, because the populations all get back together, and presumably interbreed, at nesting time."

Carr could not know for certain if the male turtles traveling to participate in the frothy gymkhana off Tortuguero from, say, Nicaragua or Colombia, arrived only to mate there with females from back home. "It will be a long time before we find out, too, because the only way a Colombian turtle is ever recognizable as such is by being picked up back home in Colombia." Carr was also aware that tagging didn't necessarily prove long-distance breeding migrations. To do that, one would have to tag at Tortuguero; then catch, record, and release where the turtle resided the rest of the year; and then catch and record once more at the Turtle Bogue. This was impossible without spending millions for boats and gear and enlisting a large corps of researchers. All the tagging project could verify was that turtles tagged at Tortuguero were later caught a long way away, and that others returned two or three years later to nest. The necessity of logically reversing all data was "a fairly frustrating way to study an animal, groping backward through its life," Carr admitted. "But it produces information, slowly. Anyway it is the only way there is."

The most important factor in the great turtle-tagging project was the human factor: the tags had to be recovered and returned by other turtle men—not other scientists but those who found tags while hunting green turtles with nets or harpoons, or with their bare hands. This brought Carr into contact with the sequestered back-country peoples of the still remote and exotic shores and islands of the western Caribbean, nowadays known as the Caribbean Rim. They are among the most ethnically diverse human populations anywhere—descendants of slaves and runaways, aborigines, colonial exiles, soldiers, merchant-adventurers, marooned sailors and pirates, from every country that once vied for power and profits on the Spanish Main—often all mixed together.

The peoples of the Caribbean are also among the most charming of Americans, and Archie Carr took great delight in the many letters accompanying the tags he received over the years, describing the capture and asking for their prize—letters often addressed to an imagined Señor "Reward Premio," which were the only two words besides the address for which there was space on the metal tags. Carr collected some of these beguiling letters in a chapter of *So Excellent a Fishe*. No one who has read them could possibly resist quoting from several, if only to give the flavor of an American dialect still spoken, in which you can hear the clashing of swords and the radical egalitarian manners of the Age of Discovery:

"After saluting you affectionately I permit myself to communicate to you having captured a turtle with a metallic plaque."

"Attentatively I direct myself to you to send you a plaque that I found encrusted on a turtle that I caught on the coast of Cojoro Venezuela. This demands that it be sent to that Institution in order to receive a reward. Receive a cordial embrace."

"Notice. This reward was fond in Colorado Bay, I'am asking, please to mail reward to George Davis in Siquirres Costa Rica. Many tanks."

Carr found his correspondents not only irresistible and essential to his reconnaissance but a lively reminder that anyone interested in the natural history of sea turtles must also weigh the human, economic, and geographic factors that constitute their unnatural history. The bulk of tag recoveries were concentrated from along the Mosquito Coast of Nicaragua, between Puerto Cabezas and the Mosquito Cays, which begins about 150 miles north of Tortuguero and continues north for roughly another 200 miles. There, growing thickly on the continental shelf, is one of the most extensive pastures of turtle grasses (*Thalassia testudinum*) in the world. The green turtles that graze there are at the heart of a special kind of coastal culture, whose story traces much of the downward trajectory of Atlantic green turtle populations from the days when Columbus sailed the oceans.

Admiral Christopher Columbus in fact discovered a great *flota* (a fleet) of green turtles cluttering the shores and shallows of some Caribbean islands south of Cuba that he named Las Tortugas (The Turtles) in May 1503. It was a breeding aggregation fit for the *Guinness Book of Records*: there were enough sexually preoccupied green turtles there to feed several growing and hungry empires. But the turtles at these islands (whose name was changed to the Caymans by Ponce de León) were not alone. Ships of half a dozen European nations scouring the neotropical shores also found them at Bermuda, through the Bahamas, the Greater Antilles, and on both coasts of the Florida peninsula, where they clogged the river estuaries at mating season.

Green turtle soon became the featured dinner menu of the New World. Many a hapless crew and starving colony were saved by the succulent herbivorous reptiles, which provided not only protein and vitamins but also a layer of rich oily cartilage on the lower shell called calipee, from which a turtle soup of celestial taste could be made. Dried and salted, turtle meat became the staple rations of the most humble sailors. Boiled in water with a few herb greens and potatoes or rice, the hearty broth sustained the plantation slave. The fine soup, steaks, curries, and stews were served on the table of the plantation master. Turtles could be hauled on deck and kept alive for fresh meat with little more to maintain them than an occasional splash of salt water. With turtle provisions, many an expedition could afford to add an entire year of sailing, trading, or marauding. Shipwrecked, one could survive practically forever on turtle meat. No sentimentalist, Archie Carr ate it with good appetite.

Carr compared the green turtle's importance in the opening of the Caribbean to that of the bison in the opening of the West. Like the buffalo to the pioneers, no food source was "as good, abundant and sure as turtle." In fact, all early European activity in the New World tropics could be said to have depended on the supply of turtles—exploration, colonization, buccaneering, and naval maneuvers. The breeding aggregation at the Caymans every summer was said by old-timer turtle-fishing captains to come from many hundreds of nautical miles away. Although there is no way of knowing their number, the enormity of their host can be imagined from the fact that it took the concentrated turtle hunting of some three hundred years to finally destroy the Cayman Islands colony.

The first record of green turtles at Tortuguero came from a Dutch ship that cruised by in 1592. But the thick belt of moist forest, swamps, and marshes kept the beachfront isolated and mostly inaccessible. Probably it was English buccaneers, ever prowling to intercept the Spanish gold and silver galleons from Porto Bello and Vera Cruz, who first encountered the resident grazing herd of green turtles off the Caribbean coast of present-day Nicaragua—from whence most of Carr's tag returns would come three centuries later.

The Drakes and the Raleighs made a market. The more they came to trade for turtles, the more aborigines were seduced from the river cultures of the interior out to the malarial coastal plains to hunt turtles. The Indians hybridized with the Englishmen who stayed to trade turtles and took native wives. Bartering turtles for guns, the Miskito Indians acquired a reputation as tough traders, willing seamen, and fierce mercenary fighters.

But the green turtles of the Mosquito Coast are in a grazing, not a breeding, mode. They do not come ashore, so one must dive for them, harpoon them from a dugout canoe, or cast a net. It is not as easy as flipping over beached she-turtles and hacking them up with a machete. And so for many generations the herd survived while the famous old rookeries of the American Tropical Zone were wiped out one by one. By the nineteenth century even the unique Cayman Islands colony was at last reduced to a mere memory of its pre-Columbian abundance by the thoughtless slaughter of nesting females.

With the depletion of the largest turtle aggregation in the New World, the famed Cayman captains turned their prows west and began to cross the open Caribbean Sea to hunt turtles on the Mosquito Banks, too. A hundred years later they were still the most important export turtle hunters in the Caribbean—until banned from Nicaraguan waters in the 1960s. By that time, several turtle-packing plants in eastern Nicaragua had turned the Miskito subsistence fishermen into commercial hunters. Even such increased hunting pressure, however, was not enough seriously to threaten the green turtle's survival. As Carr pointed out, turtle hunters on the Mosquito Coast were actually striking the species at one of its most resilient points. He wrote of the green turtle's "innate biological toughness," surviving the heavy natural predation of eggs and hatchlings, and repeated long pelagic migrations, during which food deprivation must be the norm. Even the continued devastation of turtle hunting could be borne, Carr suggested, as long as Tortuguero remained free of human exploitation. It did not.

The biological background of humans as hunter-gatherers, as René Dubos and others have pointed out, predisposes us to focus on life in the short term. Throughout human history, civilizations have trashed their Edens and paved their paradises, rarely thinking of sustaining a resource unto extinction. It has made little difference whether overexploitation is accomplished by palefaces with telephones or sun-bronzed natives shooting blow-darts. What does seem to make a difference in the survival of a species is the deadly combination of three factors: human population increase; a development in technology that increases hunting productivity; and market demand—especially the conversion of a subsistence to a cash economy, as took place among turtle hunters in the Mosquitia. Notice that none of these factors has to do with the actual population biology of the animal or resource at risk, which is wholly determined by evolution.

An example was the overexploitation of hawksbill turtles in the Caribbean in the 1960s. Japan had recovered from World War II, and its demand for turtle products, especially tortoise shell, quickly outstripped the Pacific turtle populations. Tortoise shell can be obtained only from the scutes of the hawksbill turtle. "When word got around fishing villages that a Japanese buyer would pay a couple hundred for a scute, the locals tried to kill every hawksbill they could find," said Carr's former student Larry Ogren, the National Marine Fisheries Service's turtle program director. "For the fishermen it was money in the bank. All the buyers were interested in was the shell, so they didn't have to keep the turtle or prepare it in any way. The meat could rot, for all they cared. Sometimes they just cut the shell away from the living animal and left it to die." The Japanese yen for real tortoise shell led to the vigorous slaughter of hawksbills throughout the Caribbean.

In the years after World War II these same conditions applied to the green turtles that commuted between the underwater grazing pastures of the Mosquito Banks and the Tortuguero breeding grounds. The wild beaches of the Caribbean Rim were shrinking and the human population proliferating. Technology, in the form of outboard motors fitted on dugouts and skiffs, opened a new distribution system, driving a market for eggs and meat in Costa Rica. The shining corrugated metal roofs in new seaside clearings showed the advent of the *veladores*.

In *So Excellent a Fishe* Carr describes in gloomy detail the unregulated turtle turning that by the mid-1950s was decimating the green turtles. "When I first went to Tortuguero the whole rookery was being patrolled by turtle turners, one per mile throughout the season, for the whole stretch of the nesting shore [about twenty-two miles]. If a female turtle completed her nesting and got back into the sea it was only because rain, lightning, or rheumatism kept the *velador* off the beach. . . . [Female turtles] died there by the hundreds, many of them without having nested, and virtually all without having completed the season's nesting regimen. The way things were in those times, all the green turtle populations that used the Bogue as their nesting ground seemed doomed. And because the whole western Caribbean got its green turtles from Tortuguero, the survival outlook for Chelonia there seemed sorry indeed."

Carr had no intention of leaving his puzzles unfinished. He was not marking tag return sites on a large wall map of the Gulf and the Caribbean only to note that these were the last green turtles ever to navigate there. In particular, he needed to continue in the role of Señor Reward Premio, at least until he found out where turtles go and how they get there. As a field biologist, he said, his research should not only help answer certain questions about the life history of sea turtles, but, just as important, raise new questions, outline new puzzles, pose riddles no one had thought of before.

From the beginning of the tagging program, it had become increasingly clear to Carr that the herd at Tortuguero was the last important breeding aggregation left in the Caribbean. Even if the Chelonia species survived in the Americas through individuals coming ashore to breed, the *arribadas* would be finished. Mass nestings, "one of the wonders of the natural world," as Carr called them, would be gone forever.

Carr understood that for a variety of practical reasons, conservation efforts most often address individual endangered species and their habitats. But he felt a human obligation existed to preserve less widely recognized biological phenomena, like the *arribadas*. It was an example of biological order, a "scientific and aesthetic treasure of the living world." "There is no civilized way," he said, "to escape the obligation to save them."

Carr hurled himself into conservation work body and soul. He organized other turtle scientists—and some philanthropic turtle friends from Florida—into the Caribbean Conservation Corporation in 1959. Their objective was to conserve the chelonians and other sea turtles at Tortuguero, based on sound knowledge of breeding biology. "We are not at deadlock with the green turtle as we were with the buffalo," Carr argued in *The Windward Road*. "We are killing it out idly, aimlessly, with no conviction of any sort, with most of us not even aware that it is going."

The first thing to do was stop the poaching and reduce the enormous devastation of the turtles at Tortuguero from natural and human predation. Carr was a good enough cultural anthropologist to believe that merely to ban all turtle hunting, with no means of enforcement or plan to stop the *veladores*, would only result in a black market in turtle meat. Prices would skyrocket, as always with new contraband. And high prices would in turn encourage more poaching.

It made more sense to hire the *veladores* as beach guards against natural predators; both the females and hatchlings would benefit. Carr's whole idea was to wean the *veladores* away with other jobs and incentives, not with threats and punishments, and to leave open a small legal market in turtle meat, supplied by offshore turtle fishing, until the demand for turtle dwindled or could be supplied by turtle farms. The Caribbean Conservation Corporation sponsored an experimental operation in producing turtle meat from captive-raised green turtles on Grand Cayman Island.

Carr pressed his case with the Costa Rican authorities, where he found a sympathetic ear. The former president of Costa Rica, José Figueroas, and his wife Karen, came to Tortuguero at Carr's invitation to walk the beach at night and see firsthand the havoc the poachers were doing with their machetes. The chief designer of that Central American republic's remarkable national park system, Alvaro Ugalde, had worked on the beach at Tortuguero one summer, and had gone to graduate school in the United States on a Caribbean Conservation Corporation scholarship. The Costa Ricans understood better than anyone the unique and critical nature of Tortuguero.

In 1963 Costa Rica prohibited turtle turning and egg gathering but left legal a regulated offshore turtle fishery. Carr later admitted this idea was wrong. When populations approach the vanishing point, the biological value of each individual becomes too high to allow any losses. In addition, he saw that the legal fishery led every year to the turtle boat captains wanting to fish closer to shore—they worked constantly to ease regulations. The market in turtle meat, in fact, did not disappear until after Tortuguero beach and the waters offshore were made a national park in 1970 under Costa Rican law. All turtle hunting inside the park borders was prohibited, and the prohibition was enforced. In 1975 the area was extended by 18,947 hectares.

Turtle turning was finally ended at Tortuguero. Perhaps some poaching still takes place in the "slop"—those areas north or south of the actual park boundaries where an occasional female with homing problems lands. But poaching has been vastly reduced. In addition, eggs have been removed for protection. As many as 25,000 hatchlings have been incubated and released in a given season. Experiments in keeping hatchlings—to fatten them up, as it were, against predators, which is called "head starting"—has sharply increased the survival rate of newborns. In a relatively few years the Tortuguero breeding colony was restored and is in better shape today than at any time since Archie Carr first came there in the 1950s. In a world where individual action on the environment so often appears futile, Carr's perseverance and tenacity virtually singlehandedly saved a species. A government book on the national parks of Costa Rica called Carr "a strange mixture of a visionary and a realist."

For years, two stubborn puzzles continued to dog Archie Carr: Where do the posthatchling green turtles spend their Lost Year? And how do turtles navigate in the open sea, where no landmarks are available as guides? Carr turned over in his mind every possible navigational system we know about—compass sense, celestial navigation, geomagnetism, smell gradients, even the Coriolis force (the differential between the speeds at which various parts of the planet travel through space). A lot more fundamental science had to be done. Of the two, the Lost Year puzzle was more ripe for solution.

In 1972 Professor Carr started another highly sophisticated operation: releasing drift, or Nansen, bottles, offshore from Tortuguero with tag returns inside, to try to find out where the currents would carry them. Carr wanted more data about the ocean as a transport system to prove his proposed solution to the Lost Year puzzle, which he called the "sargassum raft theory."

The theory was essentially this: the posthatchling turtles continue to swim furiously until they are far enough past the predators to meet the equatorial current, running north along the edge of the continental shelf. There the turtles crawl aboard floating sargasso weeds (brown algae), which tend to form lines along the shoreward wall of the current, like taxis waiting at the curb for their tiny passengers. Once on board, the baby turtles become "planktonic," i.e., passive like plankton, giving up control over their geographic displacement. From this point they ride the sargassum rafts into the Gulf Stream, which Carr described in Melvillean tones as: "A vast planetary swirl that starts when the equatorial current and the easterly trades push water through the Yucatán Channel and pile it up in the Gulf of Mexico. The surface there rises six to eighteen inches higher than the Atlantic level and breeds the head that drives warm water clockwise around the eastern Gulf and nozzles it out through the straits of the Florida Current. This soon meets the Antilles Current, and the two now form the 'Gulf Stream' in the new strict sense, and this moves northward with an initial speed of about three knots."

Young turtle rafters caught up in the Gulf Stream would eventually get dumped into the Sargasso Sea, the tranquil center of the Atlantic Ocean's current systems, where an estimated ten million tons of sargassum weeds circle slowly. On such sargassum rafts, Carr speculated, green turtles spend their Lost Year, feeding on the abundant raft fauna.

Gradually Carr accumulated data. Not only did drift bottles catch the Gulf Stream toward the Sargasso Sea; young sea turtles did, too. At Tortuguero Carr observed them entering weed rafts. For years he relentlessly searched the beaches of Florida, collecting records of every young turtle in sargassum weeds that had been washed ashore from the Gulf Stream off Florida's Atlantic coast. He gathered information from Florida fishermen, shoreline residents, environmentalists, and scientists.

Physical oceanography has come of age since the 1960s. New techniques have revealed vast new knowledge about such phenomena as currents, driftlines (the walls of major border currents), and convergences (the horizontal collision of water bodies). Satellite and space-shuttle imagery have shown that the world's oceans teem with such fronts of all dimensions and configurations, and that they are much more numerous and long-lasting than previously thought.

Definitive sightings and photographic evidence of juvenile turtles in the Gulf Stream and Sargasso Sea were not long in coming. Not only green but loggerhead and hawksbill turtles, too, were observed spending protracted periods of their early development as part of the sargassum raft fauna. There was evidence that loggerheads may stay there more than three years. Likewise, young turtles were repeatedly observed lined up in groups along borders where different kinds of water meet.

This evidence was good enough for a field scientist. In a special technical memo written for the National Marine Fisheries Service the year before he died, Carr claimed his own reward/premio when he wrote: "The time has come for the old term 'lost year' to be replaced by the designation 'pelagic stage.'"

But this was only the good news. The bad news was that these important stations in the passive migration of young turtles also posed a profound and previously unknown hazard to them. While Carr and others were trying to learn how sea turtles use flotsam and currents, there was also a growing recognition that human-generated debris and marine pollutants also collect along such frontal drift lines. Wherever different waters meet, downwelling occurs, not only mobilizing microplankton and concentrating biological activity but accumulating man-made wastes. Especially worrisome is the petroleum and tar concentrate in such drift lines. The remarkably sophisticated adaptive behavior by which hatchling turtles survive their Lost Year seems to lead them straight into highly polluted parts of the marine environment. About half of all mature green turtles are now affected with previously rare skin tumors.

About turtle navigation in the open sea, Carr assumed first that all animals use the same composite sensory apparatus to navigate over long distances, and that evolution may also customize one sense or another according to the species' particular migratory needs. In every case of animal migration, he said, one must try to answer two questions: "how a race of animals could learn to do it?" and "how evolution could ever instill the complex instinctive equipment required to find the tiny target?" It was not so much a theory as good scientific strategy for solving a puzzle too large even for the puzzlemaster to solve alone.

From tagging, a lot has now been learned about the travel routes of Tortuguero green turtles. Carr also tagged green turtles at Ascension Island, in the central Atlantic between Africa and Brazil. Tag returns from there demonstrated that Ascension Island turtles were grazing along the Brazilian coast, migrating more than twelve hundred miles of open ocean each way. How had they learned to do that?

Carr set out a fascinating evolutionary account in a scientific paper he wrote with a geologist named Coleman, later recounted in *So Excellent a Fishe*. At first Ascension Island was a volcanic ridge in the small and shallow sea that formed as the African and South American continents started to separate. Turtles might well have made a short swim for the selective advantage of nesting on a nearby island with few predators. As the continents drifted apart, the turtles had to travel farther and farther from their foraging grounds off Brazil to the island nesting grounds. Finally they were making the longest-known commute among sea turtles. During this period, evolution would have stamped the orientation ability to follow that route into the genes and instincts of the turtles. And it would be this migratory behavior we still see today. This could explain why a green turtle in Brazil would trace such a seemingly incongruous migratory route, twelve hundred miles against the equatorial current in order to reach Ascension Island.

To explain how, Carr worked with the U.S. Navy, trying to track turtles in the open sea every which way he could—plastic floats, radio transmitters, specially tough balloons. The silicon chip technology and satellite coverage of today was not available for him. He also knew it would take a herd of physiologists many years of hard work in laboratories to understand the internal mechanisms that guide animals like sea turtles on their enormous journeys. It was still possible that turtles were responding to features of the marine environment that humans were not even aware of or could not detect.

Science has seen spectacular advances in the study of animal orientation. The compass sense has been confirmed in a host of species. Celestial navigation has been proven in birds. Salmon use a keen adaptation of the sense of smell to return to their natal rivers to spawn. But the puzzle of how sea turtles navigate across open oceans is one that Archie Carr handed down to his students and future generations of turtle people. In 1983 he wrote: "More progress has been made during the last dozen years in revealing the sensory versatility of path-finding animals than in all the previous time since von Frische discovered the sun-compass in honey bees. But, as nearly as I can see, we are no closer to explaining the island-finding capacity of open-sea migrants than we were in 1967."

Ordinarily this is where we would leave Archie Carr. But curious readers will want to know more about the turtles. In the 1983 edition of *So Excellent a Fishe*, published as *The Sea Turtle*, Carr updated the situation:

* Great advances had been made in the conservation of sea turtles. Scientists and governments had adopted a carefully planned, comprehensive conservation strategy. This was particularly important, Carr noted, in the protection of sea turtles, which were not parochial creatures but wandered and strayed off the shores and in the territorial waters of diverse countries. The chief accomplishment of the 1970s and early 1980s, Carr could report, was in protecting nesting habitats. A belated effort was undertaken to stop the decimation at Rancho Nuevo, Mexico, where the ridley *arribada* had plummeted from more than sixty thousand to around five hundred nesting turtles in 1980. Beach posses organized by unprecedented international cooperation between Mexican government agencies and American governmental and environmental groups probably saved the ridley *arribada* a minute short of midnight.

* The successful work of the Caribbean Conservation Corporation was continuing at Tortuguero, still paying large dividends in understanding the most thoroughly studied sea turtles in the world. The coming of Tortuguero National Park and a new intracoastal waterway significantly increased human access to the ever-vulnerable nesting beach, bringing egg poachers from Limón and ecotourists from as far as Switzerland and Germany. Personally Carr said he preferred that the nesting beach be put off limits to all but himself and

his colleagues, but those days were over. Change meant the addition of guides, an information center, hiking trails, observation sites, transport, and lodging. "The only thing to do is to turn this new popularity—and its economic promise for the village and the Republic—into ammunition with which to combat poaching."

At the same time Carr was inundated with requests from other neotropical countries to help set up conservation programs. He saw that the human graduates of the Tortuguero tagging program had gone on to great things in turtle conservation—Costa Ricans, Mexicans, and Americans alike. So the Caribbean Conservation Corporation established a fellowship, inviting qualified Latin American students or conservation officers to get hands-on experience with sea turtles in the tagging program. This, too, paid off handsomely in a widening circle of experienced in-country conservation biologists.

* Despite conservation gains and the promise of head-starting and transplanting babies to try to start new breeding colonies, a whole new generation of threats had appeared. First, as already noted, was ocean pollution, especially turtles ingesting plastics, oil, tar, and PCBs. By the end of the 1980s some 44 percent of leatherback turtles stranded on American shores had ingested plastics and died as a result. Second, the development of Florida and the southeastern United States beach areas for homes was rapidly destroying the central nesting range of loggerhead turtles. Finally, incidental catch by shrimp trawlers was endangering the already much-reduced Kemp's ridley. As recession set in in the Gulf Coast oil patch, more shrimp trawlers dragged larger nets for longer drag times. The result was more turtles trapped and drowned by happenstance. Archie Carr lived long enough to see the development by the National Marine Fisheries Service labs of a successful and cheap excluder device, which even slightly increased the shrimp catch. He

looked forward to "an effective acceptance campaign," first among U.S. shrimpers, then in all regions where the movements of shrimp and sea turtles overlap. He did not live long enough to witness the Gulf shrimpers in open rebellion against these devices as a perfidious big-government plot to prevent them from exercising their constitutional right to plunder the oceans—an opening shot in the ecological counterreformation under way today.

11

Margaret Mee: Fifteen Expeditions to Amazonia

MARGARET MEE, the great artist of Amazonia and its most intrepid woman traveler, came to exploration late in life. She was forty-seven years old, a British expatriate art teacher in São Paulo, Brazil, in 1956, when she packed her painter's kit and a revolver into a canvas rucksack, and asked her husband Greville to drive her and her friend Rita to the airport. An active hiker, naturalist, and painter, Margaret had invited Rita to spend their long January holiday on a journey to Amazonia, painting botanicals. It was a drizzly day with a low cloud ceiling and poor visibility at Congonhas Airport on the outskirts of São Paulo. Margaret and Rita, dressed, they thought, for the jungle in blue jeans, long-sleeved shirts, straw hats, and boots, climbed aboard the vintage twin-engine propeller-driven cargo plane. From the doorway Margaret waved a fond goodbye. The engines hummed noisily, and the plane lifted off.

"We sat spellbound, watching the softly contoured landscape slip by, and with each stop at small towns in the middle of Brazil our excitement grew," Mee wrote in the first of her meticulously recorded Amazon journals, later collected in a book, *In Search of the Flowers of the Amazon Forests*. "Eventually we reached the impressive mouth of the Amazon in time to see the sunset spreading burning gold across sky and water. So vast was the expanse of the river that we seemed to be flying over an ocean."

For Margaret Mee it was the start of fifteen expeditions she would make to Amazonia over the next thirty years to collect and paint the dazzling variety of Amazon flora, tropical nature's botanical catalog of living beauty. Her paintings combine an astonishing technical accuracy with a breathtaking feel for color, shape, texture, and rhythm. Her works bring the individual plant and its flowers out from the thick mass of jungle vegetation. Hers was an intrepid artistic genius, brilliant in output, never before equalled, and not likely to be surpassed.

The flower-clad forests of the Amazon and its tributaries, particularly the Rio Negro, moved Mee's spirit, called to her, enchanted and haunted her. The Portuguese word for such possession by a place is *saudades*, a kind of romantic and sensual yearning which has the virtue of affecting one both on site and far removed, both distant in time as well as in the actual moment. *Saudades* is more encompassing than the English notion of nostalgia, in which the feeling is directed toward something already lost in the mists of the past. You can experience *saudades* as much gliding through a jungle, feeling the last exhalations of the heat of the day, as you can twenty years later, melancholy over the memory of that humid moment. Margaret Mee had the greatest *saudades* for Amazonia, and she returned again and again, despite the obstacles of rain and winds, floods, insects, sunburn, fevers, infections, human violence, hunger, snakes, fires, accidents, age, and just plain getting lost.

Born in England in 1909, coincidentally during the final stage of the Amazon rubber boom, Mee undertook her adventures during the Amazon's second period of uncontrolled growth, when the building of the Transamazonian Highway and similar new roads opened the fabulous emerald jewel of the world's forests to cattle barons and logging and prospecting companies, and to settlement by several million landless peasants from other parts of Brazil. The nineteenth-century rubber boom had brought sparse human settlement to the Amazon Valley, as the work of collecting rubber was done by natives and was effectively limited to regions buffering the transport concourse of the river and its lower tributaries. The second boom has proven geographically unlimited. The bulldozer and the chainsaw initiated in the 1960s the wholesale conversion of Amazon forests to settlements, ranches, and hardscrabble subsistence farms, irrespective of location.

It is unlikely, as Claude Lévi-Strauss has pointed out, that the forests bordering the thousands of miles of the Amazons were virginal even when the Europeans found Brazil in the sixteenth century. Archaeological evidence has shown that the Amazon Valley was the center of a great pre-Columbian agricultural civilization, perhaps precursor to the Incas, which vanished in the conflicts and epidemics of the Age of Conquest. If so, the riparian forests would have been only about three hundred years old in the exploitive heyday of the rubber boom, when the latex of the *Hevea brasiliensis* tree, vulcanized according to Charles Goodyear's formula, first put the world's automobile industry on the road.

The rubber boom in the Amazon Valley involved tap-

Margaret Mee painted the dazzling variety of Amazon flora during fifteen expeditions to the region's forests.

ping trees in mature forests. Rubber trees were first tapped when twenty-five years old and continued to yield latex well past one hundred years. But they were not cut down; the forest cover remained intact. This is not to say there was no development in the river system's transport corridor. The boom capital of Manaus, in particular, burst out of the jungle like a strange tumor. But after the rubber bust, jungle reclaimed much of what had been disturbed.

With Brazil's road-building binge in the 1960s, human invasion of the interior began in earnest. Whole swaths were barbered by burning and clear-cutting, exposing the Amazon Valley's poor soils to the devastating effects of erosion. As she witnessed the deconstruction of tropical nature, Margaret Mee understood that her own project was changing, too: from being the first to portray many of Amazonia's lush floral riches to very likely being the last. In her quest to make a scientific and artistic record of Amazon plants in flower before many of them ceased to exist, Mee used the power of her art to awaken people in Brazil and elsewhere to the value of the Amazon rain forests, even as they continued to disappear.

Margaret and Rita arrived first in Belém, the port city at the mouth of the Amazon's main stream, which had grown over three centuries into the great river's gateway. The water's edge bustled with motley river boats, and markets bulged with fish and fruits. The cafés served refreshing tropical fruit juices made of custard apple and passion fruit. The two women were drenched with sweat, having already discovered that white cotton clothes breathe better in the tropics than tightly woven, dark-colored denim. Margaret painted the flowers of a cannonball tree (*Couroupia guianensis*), cut for her by the gracious curator of the local Goeldi Museum. She studied the museum's collection of dried plants. A government botanist there asked them to travel up the Rio Gurupi, one of the Amazon's many tributaries, to find a strychnos vine, which he thought grew in the forests there. Margaret felt honored by the commission. When the Goeldi Museum anthropologist offered them the use of a thatched hut in an Indian village, their minds were made up. They embarked upstream, headed for the Rio Gurupi in the forests of the Tembé and Urubú tribes.

"It was a rough little river boat built of wood, with a closed cabin packed with at least twenty passengers, all from the locality and with a number of children. Conditions were far from ideal, and when we arrived in the choppy waters of the open sea children began crying and fretting as some of the women became hysterical." Thus Margaret recorded her first passage on hundreds of such small river boats, the water taxis and cargo freighters of the Amazons, with cutthroat or saintly pilots, leaking bilges and broken-down motors, gruff men offering money for sex or drunken women throwing up all night on her hammock, or halfbreeds down with dysentery, dying of tuberculosis, yellow with hepatitis.

The freelance traders who supplied the riverside population with everything from buttons to beans proved almost as colorful as the flowers Margaret came to paint. "When the motor failed (as it frequently did) or was too weak to pass a turbulent rapid, Malaki would stamp and dance, shouting, '*Vai, motor, Vai!*' (Go, motor, go!). At times he jumped lithely into the foaming river, roping the tilting craft to a rock, and then he and the boy would haul it through the rapids to calmer waters. This was the pattern of our voyage."

The two "foreign ladies," as they were respectfully called in the riverside hamlets of clustered palm-thatch huts, spent several nights sleeping in their hammocks, stretched between trees or under the sunroof of the boat deck. "I could not sleep for listening to the magic sounds of the forest," Margaret wrote. "Only the trees were silent, while the lake was alive with sparkling, splashing fishes, and the frogs' chorus mingled with the plaintive cries of night birds."

Then they transferred to a canoe, poled up the small *igarapés* (canals), "shady and filled with a rich flora," where the women bathed out of the rough currents and collected water for boiling. They camped in a primitive thatch hut with scores of potentially venomous spiders hunting overhead. In the morning a procession of African spiritist cult priestesses appeared out of nowhere and mistook Margaret and Rita for evil spirits. They continued upriver until they reached their borrowed base camp in the forest around Murutucum, and were deposited with a large bag of rice and a bunch of plantains, which they unfortunately mistook for bananas (the plantain must be cooked; it never ripens to raw edibility). Consequently they were soon weak with hunger, and Rita fell ill with fever.

Nonetheless Margaret set to work, spending her days roaming in the forest, sometimes walking many miles to find new flowers. A pair of boat-billed toucans laughed in the branches above "as we struggled to reach a white flowered orchid, a *sobralia*. The gorgeous plant had been dislodged by the previous night's heavy rain and dangled on a liana among a sodden tangle of branches. There were flowers to paint in plenty. . . . Everywhere we looked in the humid forest we found bromeliads, mostly growing on trees or branches as if stuck there by magic."

To the observer from temperate climes, where epiphytes are conspicuous by their absence from forests (except for Spanish moss in the southern United States), the dynamic and compact air plants are among the most impressive members of tropical forest galleries. Such small herbs and shrubs gain their entire necessities from the sun, rain, and tiny bits of litter on the branch; yet they constitute a surprisingly large part of the tropical forest's biomass, considering that they do not produce woody stems and branches. Among epiphytes, the bromeliad and orchid families make up the vast majority in Neotropical forests, dazzling in their amazing floral designs and intoxicating perfumes. There are more than 7,500 known species of orchids. Fifteen thousand may exist. Brazil claims more than 650 species of bromeliads, perennial plants that bloom once a year with a few dark green, leathery leaves, highly resistant to dry-season desiccation and unappetizing to herbivores. Information is sparse on bromeliads' longevity in nature, though both orchids and bromeliads have been observed to survive for tens of years if not knocked down in a storm. Dependent on water and light, epiphytes are most commonly found at higher elevations in the canopy, especially where clouds press against hillsides most of the day.

Mee sought for her paintings the living plant in bloom, at its sweetest and most resplendent—and biologically significant—moment. Many opened their flowers for only a few days a year, or a single day—or even a single night. No matter where Margaret Mee ventured in a river basin the size of the continental United States, her method of illustration required acute knowledge in order to find plants at just the right time of the year; great patience to await the precise moment of flowering; and sure-handedness to complete her drawing often against an imminent natural deadline. Yet her precise paintings tell each plant species' biological story, widening our comprehension of the term "biodiversity." Her friend, the Brazilian landscape architect Roberto Burle Marx, said of Margaret Mee, "To seek out a plant, bring it from its obscurity and reveal it to those who are inspired by Nature, is a true discovery. Each plant is a mystery whose laws challenge us; whose reason for life, whose preferences, dislikes and interrelationships, teach us a lesson which should give us a better understanding of the world in which we live."

In the forest near Murutucum, the mysteries bloomed all round her. Luckily the starving Margaret and the feverish Rita came under the protection of Antonio Carvallo, "the sage of the river." In almost every district she later visited, Margaret would find such local head men, sometimes still an Indian *tushaua*, sometimes a *caboclo* (person of mixed Indian-Portuguese descent), often a feudal "river king" of the type portrayed in Brazilian novels and movies: a tyrant, "owner" of one or more tributaries, intimidating the peasants or enslaving the Indians.

Carvallo was one of the better kind. Margaret recorded, "The peasants living up and down the river came to him to have their letters read or written or to hear news from the papers which he received from time to time. His *sitio* [ranch] was flourishing for he cultivated wisely. Instead of cutting and burning the forest around his hut, as others did, he cleared some of the undergrowth from under the massive forest trees and planted young fruit trees, using the leaf mould which the natural cycle continually renewed. His citrus, mango and native fruit trees and coffee bushes thrived under their forest protection."

Carvallo was best known for curing sick people with the medicinal plants of the forest. But he was also superstitious and subscribed to the local beliefs of a spiritful land. He feared for the women's safety, especially Rita, who had one blue eye and one brown. Carvallo worried that they would meet up in the forest with some of the descendants of African slaves who had escaped over a hundred years ago and lived in isolation, never mixing with outsiders, in the jungles of Pará State. They would think they had met the evil eye.

The two women were long overdue at their school jobs anyway. Already experiencing *saudades do Pará*, Margaret and Rita headed back downstream on another "wooden passenger-cum-cargo river boat . . . crowded with caboclos and loaded to capacity with highly inflammable *marva* [fiber similar to jute]."

On this journey, however, their luck did not hold. "The 'commandante' of the boat was fond of the bottle, and by our third night had had more than enough to drink. In this condition he insisted, against the advice of the crew, on sailing through the night. The river was pitch black and the sky starless. Soon after midnight a horrible jarring crash followed by screaming and shouting announced that we had hit a rock." They refloated the boat next day, but soon the drinking water and food ran out. The river was so low that the passengers, including Margaret and Rita, had to get out and haul the boat through the mud with steel chains. In Braganca, where they changed boats, a "sorry, squealing" herd of filthy pigs got on board. Margaret and Rita had to live in close quarters with their smell, excrement, and awful protests for the rest of the voyage. "Worse still were the ticks which deserted the dried-up creatures for us."

Mee let none of it bother her. She watched in amusement as one of the pigs sprang through the window of the hold and swam the river, chased by two crewmen in a canoe madly waving machetes, before escaping into thickets on the bank. She was hooked: "Now Amazonia was in my blood. Murutucum, that haven of green peace, seemed to be the other side of the planet, but I knew I would return."

For the next twenty-five years Mee concentrated her searches on new and unpainted varieties of flowering plants and became familiar with many types of distinct Amazonian habitat. In the forested tablelands of Mato Grosso State (literally, "denser woods"), a *cerrado*—that is, a forest of stunted, twisted trees surrounded by high grass—closed around her. The Amazon River islands, she wrote, "are paradise. I have seen scores of herons in small bushes almost submerged by water, minute purple and yellow flowers on the exposed sand, shrubs full of berries and lichens amongst which grow orchids and tillandsias, delicately leaved bromeliads. And I have found the aggressively-armed with black thorns *Aechmea tocantina*, another bromeliad."

Beginning with her third journey in 1964, Mee made the thirteen-hundred-mile tea-colored Rio Negro, reaching to the northern mountains where the Negro shares headwaters with the Orinoco River, the pivot of her explorations. Here she entered the brilliant green riparian *cataanga*—another distinctive forest growing mostly on white sand lacking nutrients. "The trees were shorter," she reported, "very green and festooned with epiphytes growing down their ribbed trunks, over the arched roots and down to the fern-covered ground. . . . It could have been some exotic hothouse." Another sand forest was "one of the most beautiful I have ever seen. Epiphytes clung to the trees half enveloped in richly green wet mosses, and in the middle of a lake surrounded by swamp grew a clusia tree bearing panicles of white flowers. It was gorgeous. The flowers were red at the centres and pendant between large oval leaves. It was just possible to reach by walking along a fallen trunk, and so Vicente was able to get me a few clusters of flowers and leaves."

Perhaps the most fertile field for her work lay in the seasonally covered areas of the river system's floodplain, called the *igapós*. These flooded forests are swamped in high water for as much as six or nine months of the year, the water covering as much as fifteen or twenty miles beyond the banks. Adverse to perennial shrubs, the *igapós* are filled with canopy trees, rooted by massive buttresses to withstand moving water. Maneuvering with a canoe in the spreading lakes of the *igapós*, Mee was literally raised up on liquid scaffolding into the canopy. She wrote, "The floods were still high, nearly thirty feet above low water, and the trees stood deep, bringing me nearer to the flowers in the canopies. I could reach to massive philodendrons crowning most of the older trees. . . . A beautifully woven nest of a japim [oriole-like oropendola] hung from a low branch. Parrots and toucans were playing and feeding in the spreading boughs. Delicate violet mimosa flowers fringed the river banks and mingled with yellow aquatic blossoms floating like fields of buttercups."

Mee exulted in the opportunity for an eye-level view of the jungle canopy and made glorious paintings of the many new plant species she found there. The *igapós* were in the lower Rio Negro, where the water more than once opened into a huge inland sea. "Before crossing this enormous expanse the experienced boatman looks at the sky and takes good note of the weather," she wrote, "for this passage can be dangerous for boats. Wind and blinding rain are the main dangers as the river is filled with difficult currents and rocks."

Upstream the rhythm of her journeys slowed to an undemanding and peaceful pace. "Jara palms grew in humid places along the banks, sometimes almost covered by the high water, their fibrous stems making wonderful homes for dozens of epiphytes. Groups of perfumed orchids . . . bloomed profusely. Cerise flowers of another orchid, *Cattleya violacea*, were so brilliant they gleamed in the trees beside the delicate white blooms of *Brassavola martiana*."

As her boat slipped past mile upon mile of jungle wilderness, Mee could record with satisfaction that animals still "pursued their lives as their predecessors had done before man made his appearance on our planet—or on the Amazon. It was glorious; time simply did not exist."

In the upper reaches of the Rio Negro, Mee repeatedly confronted tumultuous cataracts and waterfalls where the boats had to be hauled through raging torrents, pried off obstructions, and navigated with supreme skill, intuition, and body English. "The raging waters tried to hold us captive and Carlos ordered us to stamp and rock the boat to free her whilst he ran the motor at full throttle, and in the end we slipped by. Then another swirling eddy held us, and I watched a little tree on the bank as a marker to see if we were making headway against the current. This time the motor, working at its hardest, could not move through the overwhelming force of the falls. Carlos peered at the water and instinctively found a channel, shooting through exultantly to calmer waters."

Approaching the Venezuelan border, the forest was wondrously crisscrossed by a "veritable warren" of small black canals known as *igarapós*. This maze of narrow waterways forms the links between the Amazon and Orinoco watersheds, often connecting rivers flowing in opposite directions. It afforded Mee another special habitat for her sketchbooks. "The black water of the canal must be one of the most beautiful waterways in Amazonia. Its banks harbour a wealth of plants, among them the epiphyte *Aechmea chantinii* with vivid red bracts and silver ringed leaves, and scarlet pitcairnia (both bromeliads) massed on banks crammed with forest plants."

But time was already creeping up the river. The upper Amazons were then still populated mainly with natives now known as Yanamomo Indians, dressed in the traditional ornaments of their tribe, "two scarlet arara feathers bound on the upper arm and bunches of parrot feathers in the lobes of their ears." As late as the 1960s Mee noted that few Yanamomos spoke Portuguese. "In the village I was besieged by children who hung on my hands, hair, and shirt sleeves. They were smiling and chatting to me in the only language they knew, of which I could not understand a word. My replies in Portuguese, of which they could not understand a word, did not deter them in any way. They were the most affectionate and natural children one could wish to meet."

On the river, canoes full of smiling, waving Yanamomos passed her, making Mee realize that the Indians had had little contact with the *civilizados*. In the event, she had entered the last days of an intact Yanamomo culture: "When I was presented to *tashaua* [head man] Pedro, who spoke a little Portuguese, he held my hand and lovingly stroked my hair then took hold of the tresses, saying 'Margelete, I want to cut your hair.'"

She learned that measles and smallpox, rotten teeth, and lowered resistance from eating the white man's sugars and fats were destroying the Indians living around the Catholic missions. "Unhappily many of the children look sick and are suffering from sore eyes and bad teeth, whilst the adults seem healthy and have beautiful white teeth. It seems the older generation, living according to their traditions, have fared much better, though they too have been and are prey to the white man's diseases." During her fifth expedition, in 1967, she witnessed a measles epidemic raging through native villages. She was led from family to family, prostrate in hammocks, wasted by the disease, dreading the death of their children, mourning the loss of wives and husbands. In an Indian cemetery Mee "could feel the spirits of the Indians watching me as I walked sadly across their now desolate territory."

Traveling on the tropical frontier was impossible to a woman of lesser physical stamina and mental fortitude. Once, she had to pull a gun on a gold prospector who tried to rough her up (she noticed the villagers "treated me like a queen" after the incident). She was constantly black and blue from batterings in storm-tossed boats, or suffering from stings by scorpions and fire ants, and driven frantic with irritation by thousands of bites of pium, mucuim, horseflies, and mosquitoes. Another time she stopped to look at a plant, fell behind her healthy young Indian guides, and suddenly found herself alone. "I felt utterly helpless," she wrote. "I walked back a little, but seeing two diverging tracks could not decide which to take, so I shouted: my voice, muffled by the trees, sounded so weak that I despaired of ever being heard." Her hope was ebbing fast when she saw "with tremendous relief" an Indian girl from a mission coming toward her.

At the end of her fifth Amazonian journey, in 1968, waiting for a flight to São Paulo, Mee had the chance to see the way Manaus was growing. "The city was spreading into the forest . . . a road north was planned. Suddenly, this great Amazon forest which had captivated so many naturalists, one of the loveliest and most valuable gardens of nature, was being reduced to ashes by a gang of know-nothings. I was appalled."

By this time new road schemes were being announced almost daily by the Brazilian military regime, which had ousted the democratic government, suppressed political opposition with a dirty war, and silenced the press. Few in Brazil were in a position like Margaret Mee to speak out against deforestation and threats to the indigenous peoples. Even fewer in Brazil had a European platform. Mee's works were becoming widely known in Europe and North America. A folio edition of thirty-one of her botanical illustrations was published in 1968 as *Flowers of the Brazilian Forests*, sponsored by Prince Philip, Duke of Edinburgh. Visiting London for the gallery opening, Mee came down with infectious hepatitis, contracted on the Rio Negro. As she recuperated in hospital, the British press and public acclaimed her works and wondered at her Amazonian adventures.

Fully recovered, Mee returned to Brazil and in 1970 traveled up the main stem of the upper Amazon, known to the Portuguese as the Rio Solimões, or River of Poisons. This time she survived falling overboard into piranha-infested waters while brushing her teeth one morning. She lost all the plants she collected on this journey, crushed and doused with gasoline on an overcrowded passenger boat.

Margaret Mee was awarded a Guggenheim Fellowship in 1970 and used the funds for her next expeditions. She bought a small outboard motor of her own and hired a helper to carry it, so that she could fit the outboard to any hired local canoe and travel as she wanted. "With a paid crew, a small canoe in tow and my precious new motor," she wrote, "I set out into the River Maués, for the first time in charge of my own expedition."

She had resolved to spend the rest of her life recording and defending Amazonia. But the new roads, both those completed and those under construction, were already vectors of devastation. As she raced against time, her journeys became more disappointing. In place of green forests dripping with delicious new bromeliads and orchids, she found more and more "red deserts," where felled trees had exposed the tropical earth to rain and sun. Devoid of vegetation, dust bowls were developing. In one area she wrote, "The forest was destroyed for miles around. Burned giant trees, stripped of vegetation and epiphytes, stood gaunt and white-scarred-black on arid soil. The only area of green and living forest was being hacked down to make way for a rice plantation." Brazilian law said that for every tree cut down, another had to be replanted. But, she continued, "In this spot the law meant nothing. The trees had gone long ago and only maimed stumps remained. As I looked on this desert I tried to block from my mind a vision of the Amazon of the future."

Margaret Mee's adventures in Amazonia ended abruptly in 1988 when she was killed in an automobile accident while visiting England.

12

Alexander Skutch: A Moralist in the Jungle

IT WAS NOT YET DAWN when a shrill wailing of dogs rang out over the woods in the Valley of El General in southern Costa Rica. In the half-light a thick slab of rain clouds sliced off the eleven-thousand-foot dome of Cerro Chirripó, then continued to roll down through the valley. Below the cloud cover a patchwork of coffee and sugar plantations, maize and rice fields lay in a chilly blot of black-green. On the porch of the farmhouse of Los Cusingos, already fully dressed, stood Alexander Skutch, tropical America's most eminent ornithologist. Eyes closed and head cocked, Skutch listened intently to the violent barking with a look of deep distress. For a man in his late eighties, his features combine the great wisdom of age with childlike wonder, like old photos of American Indian chieftains. But also as in the portraits of the sachems, the hurts of human history are scored there, too: for while practically everyone has heard about the destruction of tropical forests—and a few even understand the dire consequences for all life forms—Skutch lives with it every day. If poachers are not hunting with their dogs in the 250 acres of rain forest whose protection has become a mission of Skutch's old age, timber cutters are after hardwoods, or men are slashing palm trees to get at the edible hearts, or peasant farmers are allowing their fires from adjacent fields to spread into the forest.

Twenty years ago the first sign of intruders at Los Cusingos might have sent Alex Skutch plunging into the dense thickets to oust them himself. But now he merely says, "I'm afraid we've got some poachers. It's not a very pleasant thing to wake up to. We'll see about them after breakfast. There's fresh coffee and tortillas ready."

The coffee at breakfast is made from beans grown a stone's throw away. The maize that makes the cornmeal that is turned into tortillas comes from Skutch's *milpa*. The oranges he serves for dessert are picked from a tree you can see—better yet, smell the lovely sweet fragrance of—from the dining room. For fifty years Skutch has subsisted almost entirely on what he can sow and harvest on this tropical farm while spending most of his time studying and writing about tropical birds.

More than sixty years ago, when Skutch first arrived as a young American botanist with a six-month grant to study the banana plant at the United Fruit Company's Almirante research station in Panama, the groundwork of the early tropical naturalists had already been laid. Most of the collecting was done, the species named and classified, the theory of evolution by natural selection accepted by science. Very little, however, was known of the actual life and habits of tropical plants and creatures—especially the nearly three thousand species of tropical birds. Skutch recalled, "While I lived and worked in that little cabin at Almirante in 1928, a rufous-tailed hummingbird built her nest just outside the window. Well, it was so fascinating to watch that hummingbird, I soon began to lose interest in the slides of banana leaves under the microscope. That bird really led me to all my studies of tropical bird life."

Skutch returned briefly to the United States, but in the midst of the Great Depression there was no work for a young botanist. He began to scour the libraries for information on tropical birds. "There was scarcely anything," he continued. "Again and again I found the phrase 'nidification unknown'—in other words, nesting unknown. Before long I said to myself, 'All these beautiful and interesting birds, and nobody knows how they build their nests, or raise their young. That's something worth studying.' That was the great gap I would try to fill."

With infinite patience, loving devotion, and a high tolerance for the trying and often lonely circumstances of the itinerant naturalist's life, Skutch set off on a fifteen-year odyssey of tropical bird-watching. His adventures took him back and forth several times between Guatemala and Honduras, five times to Costa Rica, again to Panama, and south to Ecuador. His freelance investigations became as methodical as his living was frugal. He would inhabit whatever crude shelter was available, earn a few dollars by collecting botanical specimens for American institutions, and travel on his trusty horse Bayon, with his machete tied to his waist and his revolver strapped to his hip. Skutch adopted an "opportunistic" approach to ornithology, studying nests as he chanced upon them. In a busy season he might have as many as twelve nests under observation at one time. But he never undertook to study a bird until he had the opportunity of watching its nest.

Alexander Skutch, consummate observer of bird behavior and tropical America's most eminent ornithologist.

"At a nest you have a kind of appointment with a bird," he explained to me when I visited him at Los Cusingos in 1983. "I found I could become so much more intimate with a nesting bird than merely seeing it flit through the foliage. But also, I just like seeing things built. I like seeing things cared for."

This affection for birds' domestic behavior, for the exquisite care birds give to their young, became the hallmark of all Skutch's ornithological writings. More than with any other naturalist, the love of his subject is palpable in the meticulous way he describes a bird at work making a nest:

"How does a bird attach a nest beneath a slippery, ribbonlike leaf strip, without any projection to hold the materials nor any place to perch while she works? Probably only a hummingbird, whose ability to hover motionless in the air and to maneuver in a narrow space exceeds that of all other feathered creatures, could build in such a situation. Although my first barbthroat's nest was nearly finished when found, I have watched the early stages of construction of similar nests. At first, the bird works wholly on the wing. She wraps strands of cobweb around the strip, or the tapering tip of a palm frond, while she hovers on rapidly beating wings and circles slowly around it, once or several times, keeping her bill pointing toward the leaf. She brings fragments of vegetable material and attaches them to the cobweb, then more cobweb to bind them firmly together. When she has accumulated enough material to form a little shelf on the underside of the leaf, she rests upon it, always facing inward toward the supporting strip, and often continuing to beat her wings into a haze, while she builds up the cup around herself. She uses her bill to tuck pieces in place and her feet to arrange those inside the cup. Sometimes a thin, wiry strand persists in sticking up above the rim, and, with her long bill, she tries again and again to make it stay down.

Like every other hummingbird that I have watched, barbthroats build their nests with no help from a mate."

Skutch never loses his joyful sense of wonder at the various ways birds attend and nurture their families. As with Waterton in his hammock and Belt on his mule, I always think of Alexander Skutch sitting perfectly still in his old cloth blind, waiting for epiphany as he stakes out the home of some rare and feckless tropical species. The sharp detail of description this method of study has allowed Skutch is truly remarkable, and another patent of Skutch's ornithological writings:

"Of the many surprising features in the life of the softwings, not the least is the way they feed their nestlings. . . . During the first few days after they hatch, their father stays most of the time in the burrow, brooding, while their mother brings all the food. She delivers it while standing in the doorway, with her head, or head and shoulders—rarely as much as half her body—inside, so that I could not see just what happened. With her mate inside, one might suppose that she passed the food to him, and he carried it down to the chamber and fed the nestlings. However, I am sure that, beginning at the age of three days, the blind, naked nestlings walk or shuffle up the inclined entrance tunnel, a distance of about fourteen to eighteen inches, and receive meals directly from their mother, for I saw food delivered so while their father was absent. And certain observations lead me to believe that they come to the doorway for their food from the day they hatch. When the nestlings were a day old, their mother took a minute or so to deliver a small object, which the young evidently had difficulty swallowing, although their father could have received it in a trice. On the same day, she brought a very large insect, tried for a minute or two to deliver it, then carried it away. Her mate would almost certainly have received it, had he been at the doorway. And, as I have clearly seen, White-fronted Nunbirds, still naked and sightless, toddle the length of their much longer burrows to take food at the entrance from the parents and their helpers."

Skutch's nest-watching habits led him in an interesting direction when he observed previously unrecorded instances of two females of the same species laying in the same nest and sharing feeding duties, which he described first for band-tailed barbthroat hummingbirds. He has also gathered a book of instances of fledged young or yearlings of bird species remaining with their parents to help them raise their parents' subsequent brood. With unabashed delight, Skutch described for science some of the first instances among tropical birds of what came to be called "cooperative" or "collaborative nesting," but which Skutch more modestly calls "helpers at the nest":

"It is not unusual for parent tanagers of the genus *Tangara* to receive help while caring for their offspring. In the Golden-masked Tanager, the helper may be either a juvenile from an earlier brood of the same season, readily distinguished by its full plumage, or an older bird indistinguishable from the parents. I once watched four Plain-colored Tanagers—Cinderellas of their brilliant genus—attend two nestlings. In Trinidad, four or five adult Turquoise Tanagers sometimes feed the young of a single brood."

And here he describes how the fledged females of a golden-naped woodpecker family help their parents feed a second brood: "During the nestlings' final week in the hole, their elder sisters from time to time brought them food. At least two of the adolescents, now nearly four months old, were so engaged, and probably all three were. Instead of encouraging their young helpers, the parents, particularly the mother, sometimes chased them mildly; but often she made no hostile move when she saw them approaching the nest.

"These juveniles were far from expert in their self-appointed parental role. They always came with very small particles, as though for newly hatched nestlings instead of vigorous feathered ones. Moreover, they did not know how to deliver what they brought. The parents were now feeding at the doorway, inclining their heads sideward to facilitate the transfer of the food to the nestlings' grasping mouths. The young helpers, who had not learned this trick, hesitated to approach those rudely snapping bills. Sometimes they carried the food away again. More often, after several timid advances and retreats, they gathered courage to push past the importunate nestling in the doorway. They entered with head drawn in, and probably also closed eyes, in the attitude of a man trying to shield his face from blows. Apparently, the sisters delivered the food inside the hole, as the parents no longer did.

"These woodpeckers are not the only juvenile birds that I have known to attend siblings a few months younger than themselves. I have watched Groove-billed Anis, Southern House-Wrens, and Golden-masked Tanagers feed their parents' later broods. Other naturalists have recorded such precocious parental activity in birds as diverse as gallinules, pigeons, swallows, bluebirds, cardinals, and many others."

Skutch's admiration for the highly developed family life of his bird subjects grew so intense that when he repeatedly witnessed snakes destroying nests and preying upon young birds, he instinctively intervened to protect them by killing the serpent. This broke the private vow he had taken not to kill animals in pursuit of scientific knowledge, and launched Skutch on a voyage of ethical discovery into an area few other naturalists have dared to go. By what authority could he justify killing a snake in order to let a nest of young birds survive? What moral principles guide humans in their dealings with animals?

Skutch found five worthy of consideration:

(1) Regard for human interests only. This has been the

principle most often regulating man's treatment of animals in the Western world. It is based on man's unlimited right to exploit animals simply by his power to do so. Those who accept this doctrine hold that all nonhuman creatures may be used for human ends, regardless of the suffering, destruction, or even extermination of whole species. Human interest is almost always defined here as economic interest, which raises two interesting problems with this principle: one, that human economic interests often contradict human scientific, aesthetic, and even survival interests; and two, that short-term economic interests often contradict other long-term economic interests.

(2) The principle of laissez-faire—"neither persecute nor pet." Under this rule we should permit free creatures to work out their own destinies with a minimum of human interference, not favoring one more than another of the antagonists in nature's fierce and frequent struggles for survival. Using this principle as a guide, it would be wrong to kill a snake in order to protect a bird's nest. This is the mainstream morality of the American conservation movement, where the idea is to set wilderness areas apart in national parks or wildlife refuges—to define areas where nature may reign. It has always conflicted philosophically with scientific game management and land management.

(3) The principle of *ahimsa*. This principle for regulation of human conduct toward animals comes from the East. The ancient Sanskrit word *ahimsa*, meaning "without harm," comes from the Bhagavad-Gita, where it is held that to refrain from harming all living creatures is indispensable to the attainment of spiritual enlightenment.

(4) The principle of favoring the highest. According to this principle, we take the part of those creatures we believe to be "higher" against those we hold to be "lower" on the scale of life. This "highness," Skutch says, may consist in greater similarity to ourselves, which in evolutionary terms

implies closer genetic relationship. So we would defend humans against other animals, mammals against fishes, and birds against serpents. "Highness" may also be defined as closer to humans in behavior—that is, as constituting more noble or admirable behavior, such as nurturing or cooperation—or we might believe that the more intelligent animals most deserve our protection. Skutch points out, however, that the most intelligent animals are often the ones we compete with for food and space—wolves, grizzly bears, crows, coyotes. Following the principle of favoring the highest often contradicts our economic interests and the principle of human use.

(5) The principle of harmonious association. According to this doctrine we view ourselves at the center of a nexus of animate relations, where each member of the community of life surrounding us is roughly compatible with every other, and peaceful harmony reigns around us. The garden we create or the small farm we run are examples. We exchange mutual benefits with the animals we invite there—for example, the horse works for us as a means of transport, in return for which we provide food and pasture, shelter from the weather, and veterinary care. But then a hawk terrorizes our chickens, or a snake invades the birds' nests in our garden, or a marauding cat disturbs the concord. Are we not then justified in removing the one or two that disrupt the harmony of the many?

In view of the great diversity of situations encountered by those who live close to nature, Skutch finds it hardly possible to choose a single practical principle sufficiently general to guide our relations with all the diverse forms of animal and vegetable life. He considers the principle of laissez-faire best for large tracts of forests or wilderness, where tampering with the crude balance of nature courts unpremeditated cruelty and disaster; but he clearly feels that *ahimsa*, the principle of harmlessness, is the most spiritually satisfying. In practice, however, when it comes to snakes destroying birds' nests, Skutch chooses a compromise between the principles of favoring the highest and harmonious association.

In our interview he said, "To me, something that takes care of something else is higher in a moral sense, and maybe in an evolutionary sense, too, than a creature that doesn't care about anything. Birds build their nests with exquisite care and are devoted to their young. I've seen birds welcome a young one that has just flown for the first time, and they start singing, as though jubilating, congratulating themselves and their youngster. Things like that make me feel that birds must have a fairly advanced psychic life. But look at a snake: what can it do?"

Skutch elaborates his view of snakes in *A Naturalist on a Tropical Farm*: "Is it because a snake has so few attributes of animality that it hardly seems to be an animal, but rather a creature of a unique category, as though a length of vine became able to creep rapidly through the herbage and slither up trees? It is the only large, widespread terrestrial animal that moves without limbs. It has no evident ears and cannot close its lidless eyes. Its only sound is a hiss, and, although sometimes gregarious, as when many mass together in winter torpidity, it is never really social. With few exceptions, including certain pythons, it is devoid of parental solicitude, never caring for its young.

"Many mammals and birds are likewise inveterate predators; but, by attachment to their mates, devotion to their young, a more or less developed social life, and often, too, certain indications of playfulness and joy in living, they may stir our sympathy. The serpent is stark predation, the predatory existence in its baldest, least mitigated form. It might be characterized as an elongated, distensible stomach, with the minimum of accessories needed to fill and propagate this maw—not even teeth that can tear its food. It crams itself with animal life that is often warm and vibrant, to prolong an existence in which we detect no joy and no emotion. It reveals the depth to which evolution can sink when it takes the downward path and strips animals to the irreducible minimum able to perpetuate a predatory life in its naked horror. The contemplation of such an existence has a horrid fascination for the human mind and distresses a sensitive spirit."

In the solitude of tropical forests, Skutch derived a highly personal and spiritual, quasi-oriental view of natural evolution, with a dark side of predation, competition, violence, and strife, and a brighter side of cooperation and harmony. "Perhaps one of the best developed forms of cooperation is the flowering and fruiting of trees," he explains. "When a tree flowers, it attracts pollen-gathering bees, then it sets fruits or berries. The birds come and eat them, later passing them through the alimentary canal and spreading the seeds, from which grow more trees and shrubs to provide more pollen for industrious bees. That forms a peacefully cooperating community. The bees don't fight among themselves. The birds get along very well. So that's a very advanced form, where dozens, or even hundreds, of different kinds of organisms cooperate to keep the cycle going. This benign cycle, in which every participant is benefited and none is harmed, is one of evolution's finest accomplishments, proof that a blind, undirected process, which depends upon random variations and produces much that we abhor, and much that we regard with mixed feelings, can also create much that we unreservedly applaud."

Through the 1930s Skutch's name began appearing in scientific journals as the author of new life histories of a myriad of tropical birds—black-chinned jacamars, slaty antshrikes, northern tody flycatchers, and Reiffer's hummingbirds, to name only a few of hundreds. Finally, in 1940, with World War II under way in Europe, the U.S. Department of Agriculture commissioned Skutch to survey the Peruvian Amazon for its potential to grow strategic rubber. Skutch seized the modest grubstake earned on the expedition and returned to the place he had decided was "a naturalist's paradise"—the Valley of El General in the southernmost Costa Rican province of San Isidro del General. Skutch was intent on purchasing land, building a home and library, surviving on subsistence farming, and continuing his ornithological work in a more settled condition. The land was cheap and the setting perfect for his purposes. A high range of mountains, the Cordillera de Talamanca, separates the rain forests of the Caribbean lowlands to the northeast from the lighter woodlands of the Pacific lowlands to the west, which endure longer dry seasons. Eastward lie the savannas and gallery forests of Panama's Pacific slope. In this location Skutch found montane or elevated rain forest with more distinct dry-wet seasons—better for farming—a cooler, less humid climate, and an avifauna combining transition-zone species with unique species, including the turquoise cotinga, fiery-billed aracari, Baird's trogon, golden-naped woodpecker, white-crested coquette hummingbird, and other splendid, little-known birds.

Soon books and articles started flowing from Skutch's temperate pen with all the profusion of the tropical environment. He specialized in old-fashioned life histories—detailed studies of selected species, reporting their life cycles, how they grow and mate and reproduce, what they eat, their relations to the total environment, physical and biotic. This is the kind of time-consuming, descriptive field biology increasingly shunted aside as technical science turns to number-crunching. In 1954, at age fifty, Skutch published the first volume of his magnum opus, *The Life Histories of Central American Birds*. When volumes two and three appeared, along with *Life Histories of Central American Highland Birds*, in the 1960s, Skutch's place in the front rank of

Neotropical ornithology was assured. As Roger Tory Peterson observed, Alexander Skutch did for tropical birds what Audubon had done for the birds of North America. One is tempted to add: with a lot fewer casualties.

The wanderer had also fulfilled his dream of home. In 1950 Skutch married the former Pamela Lankester, daughter of an English coffee planter long established in Costa Rica. In 1968 they adopted a son, Edwin. Meanwhile, not satisfied with writing only for a small professional audience, Skutch sought a more popular form of expression for his decades of travel, observation, and reflections. He found an eclectic style of autobiographical memoir, filled with lyrical scenes of tropical America, unforgettable descriptions and incidents of bird life, and wonderfully lucid, if idiosyncratic, moral and philosophical essays on nature. Skutch now has more than twenty books to his credit, but perhaps none would have been written had he not put down roots at Los Cusingos. As he writes in *A Naturalist on a Tropical Farm*, in words that could put moss on a rolling stone: "Under his own roof, with his own land and trees, his own animals and stored grains and crops growing around him, one feels somewhat sheltered from the chill vacuity of interstellar spaces, and even from the violent social agitations that torment mankind today. He can go about his work with unhurried deliberation, a sense of ample time for its completion, for which the rootless wanderer, with neither a home nor sustaining reserves, yearns in vain."

After breakfast that day, Skutch strapped his old machete round his waist, and we set off down the trail into the forest. The dogs had vanished, and the sun obliged us, piercing the foliage in cathedrals of light. It was easy to feel daunted in the presence of one of the world's great tropical birdmen, but Skutch proved as amiable a trail companion as he was an enthusiastic teacher. Nature seemed to blush and tell her secrets at his approach. Stooping down, he pointed out the artfully camouflaged bundle of twigs and grass that wrens used as a sleeping dormitory. A few steps farther, Skutch picked up the tiny, curved inflorescence of a wild plantain and explained how certain hummingbirds had evolved hooked bills, precisely designed to reach into the plantain's deepest recess for nectar and insects. Nearby a red-capped manikin flitted onto a branch, brilliantly black all over with a fiery red head patch. Skutch described how male manikins, like hummingbirds, gather in "singing assemblies," advertising themselves to female manikins with eggs to fertilize. "Each male manikin tries to attract females to himself," he said. "Yet they're all there together, in a conspicuous place that persists from year to year. So they're competing and cooperating at the same time, and mostly they do so peacefully. I've often tried to point out that these two aspects of nature—cooperation and competition, are sometimes combined in the most unexpected ways. If you overemphasize one or the other, you get a distorted view of nature."

From time to time Skutch allows touring bird-watching groups to visit Los Cusingos. He has also conducted classes in field ornithology from his front porch for students from the University of Costa Rica. He constantly stresses to visitors how everything fine that nature has created for humans to appreciate in the tropics is vanishing swiftly under the relentless pressures of overpopulation, poaching, and the sacrifice of patrimony for economic gain.

"When I first rode into this valley there were great swaths of forest everywhere," Skutch told me when we returned to the farmhouse. "Most of it has literally gone up in smoke, for pasture or coffee plantation. I decided quite a while ago that for a small population, slash-and-burn agriculture is probably the best system for tropical lands with high rainfall. But traditional migratory agriculture is disappearing as more intensive methods come in. For example, the coffee plantations used to have tall shade trees, especially leguminous trees, which added a lot of nitrogen to the soil. Some of my best birding has been in coffee plantations with those big old shade trees. But coffee under shade doesn't produce as heavily as coffee under sun. Now they plant a new variety of low bush coffee, without shade. They get a bigger crop, but they have to fertilize heavily, and the open country birds disappear.

"Sugar, too, has changed. Plantations used to run their own small mills and cut cane selectively. Now they plant big fields and clear-cut—not as good for the birds. And don't even get me started on the subject of beef cattle. Throughout tropical America the forests are being destroyed at a terrific rate to raise beef cattle for export. All those fine forests. And for what? They sell the meat to the States, where Americans eat too much meat anyway, and they get dollars which go right back to the States and Japan, mostly for a lot of electronic trash. It seems to me that a country that cuts its forests for beef cattle for export is sacrificing its patrimony for temporary gain."

But Skutch also has an ethical beef, as it were, with beef cattle. Much of Skutch's life has been a quiet but insistent revolt against the harsher aspects of nature: "Predation, the exploitation by one organism of another to supply its vital needs, is a major source of the ills that afflict the living world. In its subtle form of parasitism it causes prolonged suffering rather than sudden death. In its more spectacular modes, as when a lion pounces upon an antelope or a hawk strikes down a bird, it is responsible for more insidious evils. More than the occasional violence of the elements or the competition between individuals of the same species for territory, food, or mates, predation has brought fear and hatred into the world. Doubtless it is because man's ancestors were for ages not only fiercely predatory animals but also frequent victims of predation by the larger carnivores that his passions are today so violent and difficult to control, his rage so intense, his hatred so implacable, his fear so enervating. Man the omnivorous predator became man the merciless raider and warrior. The clubs and stones he employed to kill his prey slowly evolved into spears and arrows, and, finally, into artillery and atomic bombs."

When Alexander Skutch looks out over the Valley of El General and sees the green forest gradually disappear, he is acutely aware of tropical deforestation from the unique point of view of a vegetarian evolutionary biologist: "Green is the color of chlorophyll, the most beneficent, constructive substance on earth. During every daylight hour, it is silently, steadily engaged in photosynthesis, the process that supports all the life of this planet, except the minute fraction of obscure organisms that depend upon chemosynthesis. Every movement that we make, every thought that we think, every pulsation of every heart uses the energy that this wonderful substance captures from sunlight and stores in life-supporting compounds. Moreover, we owe to its ability to decompose carbon dioxide the atmospheric oxygen without which most organisms could not live. Even in the oceans, where it mostly passes unnoticed because it is distributed among minute planktonic organisms, or is masked by the browns and reds of algae, chlorophyll is present in vast amounts, synthesizing food in quantities comparable to that produced on land. How can evil predominate, how can pessimism prevail, how can a thoughtful mind sink into ultimate despair, on a planet colored green with a substance so beneficent as chlorophyll, engaged with quiet efficiency in the constructive work of photosynthesis?"

As Skutch believes that photosynthesis is the basic good of the living world, so he teaches that predation is the basic evil, the cause of most of the ills that afflict life. And as predation, parasitism, and "every form of ruthless exploitation of one organism by another" are proof of evolution's "tragic failure to create a harmonious community of living things," so he believes that every instance of fruitful cooperation is evolution's triumph—a sample of what might have occurred with a different sequence of mutations over the ages, or if the evolutionary process had been guided by what Skutch terms a "wise and compassionate Intelligence." Or what might have occurred elsewhere in the universe, on a distant planet happier than our own.

Skutch views his own revolt against nature's harshness as allied to the spiritual tradition of the East—Hinduism, Buddhism, Taoism, and especially Jainism, an ancient Asian sect now numbering two million to three million adherents in northern India, whose quest for enlightenment includes both vegetarianism as part of *ahimsa* and a conscious striving by humans to create a harmonious and happier world.

Sad to say, the intellectual courage with which Skutch has criticized the evolutionary process—criticized nature herself—has met with underwhelming skepticism from most biologists, who are uncomfortable with the cocktail of science and spiritualism Skutch serves up—and who are, by the way, mostly red-blooded carnivores. Skutch has been universally lauded for his ornithology and snubbed for his philosophy.

Yet Alexander Skutch means to challenge us and stimulate change, not threaten us. His argument is that in the absence of a benevolent intelligence to guide her, nature has depended on random mutations, blundering forward into a world of brutality and suffering—excessive multiplication of animal populations exceeding the plants' ability to sustain them, dependence on predation, and thereafter the advance of every mutation giving predators or prey an advantage in the struggle to survive—penetrating fangs, grasping talons, tearing beaks and poison glands, cryptic coloring and mimicry, along with the strength, speed, cunning, and emotions appropriate to the killer. Thus a planet made habitable and fruitful by green plants silently absorbing sunlight became the stage for carnage on an incredibly vast scale. Skutch wrote:

"When I reflect upon this tragic turn that evolution has taken, I gaze over my verdant valley with mixed feelings. The sight of countless green leaves steadily engaged in an activity wholly beneficent and constructive dispels black pessimism but not all dark misgivings. Everywhere, photosynthesis, nature's brightest aspect, is decreasing as man covers larger areas with his highways and constructions, destroys thriving forests to make cultivated fields and pastures for his beef cattle, contributes to the spread of deserts by overexploiting arid lands, and poisons seas with his wastes. Simultaneously, predation, nature's darker aspect, grows apace, as increasing areas are devoted to raising cattle for slaughter and the oceans are more thoroughly scoured for the flesh of their living inhabitants. . . .

"The exuberant tropical vegetation that I survey reminds me that nature's boundless creativity is not tempered by restraint. Its excessive multiplication of species and individuals is responsible for the gravest ills that afflict the living world, including the prevalence of predation. Unrestrained creativity is the precursor of destruction. Life would be so much more pleasant if there were fewer living things! What has been most conspicuously lacking, in the natural world as in human society, is moderation, which Plato and other Classical philosophers held to be the highest good, the key to every virtue. Moderation, which requires thoughtfulness, measure, and self-control, could be man's most important contribution to the life of this planet. Without moderation, life will never rise, upon the firm foundation of photosynthesis, to the heights that this foundation might support, nor will man realize all the splendid values within his reach. Unless we exercise restraint and moderation, in reproduction, in consumption, in our exploitation of other organisms and the demands we make upon nature's bounty, the fairest planet in the solar system will not long remain a fit abode for life."

The afternoon darkened, and the rains soon began pelting down. The downpour drove the hummingbirds from the hibiscus in Skutch's garden, and the green honeycreeper off the feeding platform Skutch sets out with bananas and sweets for his winged friends. We could not shout over the rain, so Skutch sat in his study, his eyes closed, serenely listening to the splashing music of the shower. In the simply furnished study of the unelectrified farmhouse, with no clock in view, no telephone ringing, no television, and sheaves of water dripping from the open window sashes—Skutch has never permitted glass panes or screens to form a barrier between his home and nature—one had the sense of being on a frontier. And in many ways the American tropics are like the American frontier in its closing days, when Temperate Zone forests had been cleared for farmland, Midwest marshes had been drained, and birds and mammals had disappeared along with so much wilderness. The difference is that the tropical frontier is disappearing faster because it is more heavily populated, because the pressure on land and resources is global, because the poor countries of the South have little alternative to exploiting nature according to the primacy of human economic interests, and because the technology of ecological destruction is more advanced.

When you think of Skutch riding horseback on the tropical frontier all those years, he seems a living legend—the epitome of the rugged individualist, pursuing his dream of studying tropical birds, with little to sustain him but his own inner resources. Then you think of him at Los Cusingos—this aged, convivial ascetic, communicating his vision of nature and defending singlehandedly his small forest and its animals. And you can't help being reminded of Thoreau, had he stayed at Walden, of Albert Schweitzer in Africa: like them, Alexander Skutch is a prophet, crying in a shrinking wilderness.

13

Daniel Janzen: How to Grow a Tropical Forest

ILLUMINATION. FOCUS. CLICK.

A slide appears on the screen. It shows a bare white map of Central America and Mexico, the Caribbean Sea to the east and the Pacific Ocean to the west. Running down the entire Pacific coastline is a thick green swath.

"When the Spaniards first hit the Americas in the sixteenth century," informs the nasal voice of Dr. Daniel Janzen, a tropical ecologist, University of Pennsylvania biology professor, and a kind of latter-day Darwin, "there was tropical dry forest over everything marked green on the map, an area the size of France—or five Guatemalas."

CLICK.

New slide, same map. Only now the green swath is gone. Instead there are several red dots along that same Pacific coast. The year is 1986, and moving into the beam of light to explain that the red dots are the 1 percent of tropical dry forest patches remaining, is Janzen himself. Baja, California, squiggles down his imposing forehead like Gorbachev's splotch, ending in Janzen's thick, unkempt gray beard. Somewhere off the coast of Mexico, the tails of his blue workshirt have sprung loose from his baggy khaki trousers, where a Swiss Army knife and a key ring as big as the Ritz are hitched to his belt by a tattered shoestring. Sleeves are rolled up high just around El Salvador, and his shirt pockets bulge with the naturalist's portable minilab—notebook, magnifying glass, pens, and so forth. If you dug down into the pocket litter you would undoubtedly find a few tropical seeds, picked out of a pile of mammal dung, or perhaps several tropical moths, or at least their pupae.

A murmur runs through the audience, composed of wealthy Philadelphia philanthropists and foundation and corporate donors, called together by a friend of Janzen's. They have been sipping cocktails served by college boys in black tie, and it is hard to know if the murmur is a reaction to Janzen's bad news about the loss of Central American dry forests or to the wild and woolly Janzen himself. This is not the kind of audience that usually congregates around an esoteric tropical field biologist.

Nor was Dan Janzen accustomed to these deep pockets. But he was growing more comfortable with them as he crossed the country and the world, from benefit reception to cocktail party fund-raiser, from lecture to symposium to conference, from auditorium to boardroom, in his campaign to save tropical dry forest. For one thing, he could now make it to the front of a crowded room without crushing anyone under his restless, rural stride. On the other hand, he had not yet come to the point of putting on a suit and tie. At the

Daniel Janzen's visionary plan for a national park in Costa Rica may be the most important initiative in the Americas to conserve tropical forests.

age of forty-eight he apparently did not own a tie—nor a home, a car, or a conventional family life. But that night he had replaced his usual mud-and-dung-covered flapping work boots with a pair of sensible black oxfords. And he made it through the evening without once using the word "shit," which in Janzen's case is usually not an oath but rather the no-frills term for an important part of his ecological studies, as in "tapir shit" or "agouti shit." On this evening he was also notably without the feral smell that usually clings to him as he goes about fieldwork at his research station in Santa Rosa National Park in Guanacaste Province, Costa Rica—an odor one of his colleagues dubbed *eau de peccary* in honor of the wild pigs native to Central America.

Otherwise, squinting in the light from his slide projector, he was looking very much like vintage Dan Janzen. It may be apocryphal, but the legend goes that at the University of Pennsylvania, where he usually spends the fall semester teaching tropical biology, sleeping in his office amid the whiz and whir of his trusty Macintosh, Dan Janzen has been mistaken for a slightly addled janitor. This image scarcely bothers him, for Dan Janzen is indeed the custodian of a biology institute, though not one that is part of any university, housed in a conventional building of bricks and mortar.

Janzen is custodian, architect, chief contractor, fundraiser, administrator, director of research, and head cheerleader for approximately three hundred square miles of current and former tropical dry forest that he crusaded in the 1980s to buy, to restore ecologically, and to preserve as Guanacaste National Park. Having spent the better part of three decades learning how plants and animals interact in tropical forests, and pioneered the techniques of what is called "restoration ecology," Janzen said his chunk of Costa Rica contained everything he needs literally to grow a new jungle over this ecologically "trashed" area—something that nature unaided (but unhindered) would require approximately half a millennium to accomplish.

Dan Janzen's plan for Guanacaste National Park has been called the most significant project in tropical forest conservation in the Americas. It may also be the most visionary environmental scheme of the end of the twentieth century. And Janzen may even be capable of achieving it. Already honored as one of the world's greatest living naturalists and the dean of tropical ecologists, Janzen's scientific eminence was recognized in 1984 when the Swedish Academy of Sciences awarded him the $100,000 Crafoord Prize, regarded as the equivalent of a Nobel Prize in the natural sciences. The academy cited Janzen's "imaginative and stimulating studies on coevolution"—the mutual adaptations between plant and animal species that make them dependent on each other—and called him a "superb observer." Janzen's more than 250 published studies and books; his 40-year study of the tropical forest, including 25 years of studying moths; and his projected lifetime goal of understanding the impact of several thousand herbivores on Santa Rosa's 27,000 acres of vegetation, leave no doubt that he is also a champion Clydesdale when it comes to fieldwork. His commitment to probing one territory in northwest Costa Rica is reminiscent of other scientific luminaries whose profound ideas emanated from association with one turf—Darwin with the Galápagos Islands, Linnaeus with Lapland, Gilbert White with Selborne Parish.

Behind all his brains and energy, Janzen is a splendid seat-of-the-pants naturalist who has turned his curiosity, avidity, tenacity, and highly developed powers of observation into a singular career. He casts his eye over multiple facets of the natural world. His studies bound energetically across scientific disciplines and probe restlessly into the way tropical nature works. You rarely find Janzen confining himself to one scientific field. When he observed how the spike-horned trees called ant-acacias manufacture a sugary liquid that insects can subsist on, he was simultaneously studying how a certain fiercely defensive insect called acacia-ants protect the trees in return for their sustenance and housing.

Such interactions between plants and animals are mutual adaptations. And having developed over time, they exemplify the coevolutionary process—species developing together. Janzen is master of this tropical form. His studies of tropical moths, for example, are also studies of the plants that larva moths eat; and what other species eat those plants; and what species eat the larvae; and what plants do to avoid being eaten, by making themselves bad-tasting to certain animals, too sticky to mess with, too thorny, perhaps unreachable, or, as in the case of the ant-acacias, protected by armies of stinging ants!

Janzen is an entomologist, a lepidopterist, a botantist, a herpetologist, and a paleontologist; but most significantly he is an encyclopedist of the tropical forest who thinks not only about particular species but about the larger questions of how natural systems develop and function and change in the Neotropics. When he describes a species, or interaction, or a system, Janzen sees not only what is happening in the present. He sees biological facts and events of the past. He tries to fathom how changing selective pressures led to adaptations we see now—even though the environment circumstances may have changed long ago.

Few American scientists have had the opportunity to study tropical ecosystems over such a long time as Dan Janzen. Generally they make forays into the tropics or spend summer breaks and sabbatical semesters doing field research. If they are studying seed dispersal, for example, they choose a particular tree to study, observe and record which animals pollinate its flowers and which eat its fruits, then write up the interactions and move on.

Janzen has a somewhat more than academic contempt for such "part-timers." He has a better story. Over the course of decades living five to nine months a year in Central America's wildlands, for example, Janzen observed a set of trees, including the honey locust, osage orange, pawpaw, and persimmon, which produce fruits so large or oddly shaped, oddly colored, odd-tasting, or odd-smelling, that no animals are regularly attracted to them as a food resource. Once in a great while he saw a horse or a steer, a peccary or a monkey, eating one of these strange fruits. Janzen watched, waited, and pondered.

At length he speculated in a paper, "Neotropical Anachronisms: The Fruits the Gomphotheres Ate," that if nothing alive regularly eats the fruits, something extinct must have. Before the Ice Age, Central America was populated by megafauna—brawny herbivores whose fossils Darwin collected in Patagonia: mammoths, mastodons, giant ground sloths, glyptodonts, megatheria, toxodontias, gomphotheres, and the forerunners of the modern horse. Janzen thinks the tree species that produce the odd, oversize, and uneaten fruits must have coevolved with the megafauna, which probably found them appetizing. A gomphotheres was just about the right size to eat a bowling-ball-sized, pithy fruit. In return for nutrition, the megafauna spread (and fertilized) their seeds through defecation. But as the Ice Age slid away, humans migrated south into the Americas. The megafauna declined, perhaps hunted to extinction.

The trees, however, survived the last ten thousand years as what Janzen calls "evolutionary anachronisms"—living representatives of doomed species. Lacking their mutually evolved seed-dispersal agents—the megafauna—the trees could reproduce only sporadically, randomly, on a lucky wind, or thanks to the introduction of horses and cattle. Nor was their survival ensured by monkeys or peccaries as seed dispersers, as some part-timers had suggested.

Janzen's conclusions? "If Neotropical ecologists and evolutionary biologists wish to determine who eats fruit, who carries sticky seeds, and who browses, grazes, tramples, and voids that segment of the habitat that would have been within reach of a variety of megafaunal trunks, tusks, snouts, tongues, and teeth, the missing megafauna must be considered."

Not surprising, Janzen's paper brought protests in scientific journals, especially from those who had published information on monkeys or peccaries as seed dispersers. What about statistics to back up the megafauna theory? Janzen had few to offer. His megafauna theory rests on anecdote. It was "speculation," said one letter.

"Speculation?" Janzen huffed in an interview. "Ecology is all speculation! But you better know your natural history before you propose your theory. These biologists waste their time looking at the way a particular present-day species—monkeys, for instance—interacts with a present-day species of plant, never questioning whether this plant might have evolved someplace else under different selective pressures. They look at evolution as if it began around 1940.

"But the selective pressures under which plant and animal characteristics evolved are no longer present in the American tropics," Janzen continued. "You have to look at present-day species in light of five to ten thousand years of human disturbance to the ecosystem. What we're seeing when we look at tropical forests today is not virgin, pristine, or primary forest. It's an ancient field full of successional weeds. They're just weeds that happen to be two hundred feet tall."

Convinced that he was right about the coevolution of

odd fruit-bearing trees and the gomphotheres, Janzen set about devising a way to collect data to silence the academics. His means was a small traveling circus that went from Mexico to Panama each year along the Pan American highway, which passes by Janzen's headquarters in Santa Rosa National Park. And the circus had an elephant.

The elephant is, of course, a modern-day representative of the megafauna. And Janzen, being Janzen, decided that his best strategy was to kidnap the elephant, turn it loose among his strange fruit trees in Santa Rosa, then follow it on foot, taking careful notes on what it ate and what seeds it defecated. Unfortunately for science—but fortunately for the elephant—civil war between the Sandinistas and Somoza erupted in Nicaragua, and the circus no longer passed by. Janzen settled instead for releasing horses.

Janzen's own research has taken place in tropical deciduous, or dry, forest where the rainy season–dry season pattern is strongly marked by trees dropping their leaves from December through April. Those areas of tropical America originally covered by deciduous forests happened to have richer soils and more propitious climates for agriculture. Consequently they have become the breadbaskets and plantation regions of the Neotropics. Janzen has said that if a cash crop had been found that grew as easily on rain forest soil as the cotton and cattle that grow in former dry forest areas, there would already be no tropical rain forest left to save.

By the 1980s Janzen realized that if he and other tropical biologists did not act to save tropical wilderness, there would soon be little of tropical nature left to study, other than the effects of human intervention. The 5 percent of forest coverage left in Gaunacaste was chosen by default, agriculture being the primary land use. Tropical ecosystems do not follow human land-use patterns; but when man determines where crops will be raised, he necessarily also determines where wildlands remain and where nature reserves can be established.

Janzen's moth collecting over decades at Santa Rosa showed, for example, that Guanacaste Province is a highly seasonal site for its thousands of moth species. When the trees drop their leaves in the dry season, many moths move to cooler evergreen forests at higher elevations, where they use less energy and may even reproduce. Janzen says other Santa Rosa moths probably migrate across the whole Costa Rican landmass from the Pacific to the Atlantic coast lowlands, synchronizing their migrations in order to obtain food and moisture, find mates, or lay eggs.

"Looked at the other way round," Janzen writes in his introduction to the insects in *Costa Rican Natural History*, "late in the rainy season a number of butterflies appear at Santa Rosa National Park that are thought of as being normal residents of the evergreen forests to the east. For example, throughout the year occasional morpho butterflies (*Morpho peleides*) are seen flying up a riverbed or taken in a bait trap in the park, but in November they are extremely common in riverbeds and other moist habitats."

Tropical wildlands are preserved by governments according to what lands are available *after* economically desirable lands have been cleared for agriculture or by timbering. Janzen has demonstrated that simply preserving wildlands by default may not be enough to save moths, given that the adult moths are changing habitats on a regular basis in their migratory survival strategy. They would need to have access to all the habitats they use. To create tropical reserves opportunistically, without knowing the habits and needs of the species which inhabit them, may not accomplish as much as desired in preserving biodiversity.

In most cases, claims Janzen, conservation in the tropics is not about just preserving the relict wildlands but about trying to restore already trashed habitats—a process that can only be accomplished with the help of biologists who understand what the habitats were like before they were trashed—and how to return them to that status. As it happens, Janzen's long years of research into the interactions of plants and plant eaters have taught him what he needs to know about restoring a tropical dry forest system. He knows who eats what where in the tropical dry forest of Guanacaste Province. He knows who shits what where. This is the key to what Janzen calls his "Action Plan" for growing a tropical ecosystem containing the inhabitants it originally embraced as primary forest.

"If all fire and livestock were deleted from GNP [Guanacaste National Park] today, and the site simply allowed to revert to its own vegetation, the grass patches of less than 5–10 hectares would be largely woody vegetation within 20 years while the largest expanses of pasture will require 50–200 years to attain this status," Janzen writes. "The entire area will require at least 100–1000 years to begin to approximate the full structure of pristine dry forest."

Much too long for the impatient Janzen, who says the natural forest reinvasion processes can be substantially speeded up by habitat manipulation. The first part of his Action Plan is fire control. As elsewhere in the American tropics, fire is the single biggest threat to forests. Not natural forest fires, since lightning does not occur in the dry season and Janzen has found "absolutely no circumstantial or biological evidence that natural fires were ever part of the Guanacaste dry forest environment." All fires are anthropogenic, as tropical farmers traditionally "clean" their fields and pastures by burning them off at the end of the dry season before planting. These fires can spread into forested areas downwind, hot enough to kill all above-ground small woody plants and sublethally damage larger trees.

"The technology of fire elimination is feasible and straightforward," Janzen writes. Such fires are usually extinguished by setting backfires or beating them out. Janzen has demonstrated that incoming fires can be prevented simply by burning a one-hundred-to-two-hundred-meter fire lane along all forest borders adjacent to pastures and fields. ("Ideally," he says, "the fire lanes are burned on pastureland belonging to neighbors.") When the wind-driven fire reaches the fire lane, it lacks fuel to spread, and dies. In the case of a fast-moving daytime fire burning toward the park, a backfire started from the fire lane will stop it. Fires within the park can be combated most effectively by fire crews reaching them while they are still small, and both backfiring and directly beating out the fire. "GNP fires are smoke-rich and can be located easily from a high point if a fire watch is maintained during the dry months," adds Janzen. "When the fires are stopped, the return to woody vegetation can be rapid even in the absence of livestock to depress competing grasses."

Once fires have been controlled, the "who shits what where" that Janzen has been studying for so long comes into play. A rich wild fauna in dry forests naturally moves seeds into pastures, fields, and woody succession. The first wave of tree seedlings to become established in pastures usually consists of wind-dispersed tree seeds from upwind-side forest, moved by dry-season winds. Although such seeds tend not to reach the centers of large pastures, they can easily be scattered there by hand.

Once a few such trees are established, animals crossing the pastures begin to use them. Birds and bats temporarily perch on such trees. Mammals (deer, peccaries, tapirs, coatis, cows, horses, coyotes) will often pause or pass beneath them. The animals defecate the seeds they have eaten nearby, the seeds sprout, and new seedlings take root. Soon a growing island of animal-dispersed seedlings begins to appear around the large isolated wind-dispersed trees. Once a core of new forest patch is growing, livestock help not only by spreading seeds but by suppressing competing thick grasses. The use of horses and cattle must be highly controlled and terminated as new forest patches reach the stage where grass is no longer threatening succession through competition and fire.

How long will it take Dan Janzen to grow a tropical forest? His Action Plan cannot predict. That depends on the effectiveness of fire control and seed-dispersal management. But Janzen is confident of his methodology. "We already have the biological knowledge that is needed to make GNP a reality," he declares flatly in his Action Plan.

Dan Janzen never calculated what his fund-raising crusade cost him in personal terms—comfort, affluence, social status. He calculates only time. He does not take holidays off; he does not like the idea of holidays any more than he likes the idea that he must take time to sleep and eat. At lunch in Philadelphia one day, shoveling his food in with a spoon (a fork takes too long), Janzen made a typically extraordinary statement: "I know exactly what I'll be doing every year for the rest of my life."

What he did in the short term was singlehandedly to raise $11.8 million to establish Guanacaste National Park in 1989, which accounted for his somewhat grudging forays into what the rest of us call the real world. Having planned his workload for the rest of his days, Janzen is aware that he will not be alive to see the end result of his dream of growing a tropical forest. But a little inconvenience like mortality will not stop Dan Janzen.

CLICK.

A tropical landscape. Not lush, not green, not rainy. So hot that you get up at 4 a.m. and go to sit with the toads under wet leaves. The desiccated plain in Guanacaste Province is covered with overbrowsed flaxen Jaragua grass. As far as the eye can see there is nothing that looks even remotely like a national park. High up on the partly denuded sides of the twin volcanoes Cacao and Orosi, smoke rises in acrid columns. It is March, and the farmers have been out burning last night before planting cash crops of cotton, rape, cashews, corn, beans, squash. I am looking out from Santa Rosa into the core of Janzen's dream, Guanacaste National Park.

Along the road, next to a stand of leafless and sweltering forest, comes Janzen himself, now more normally attired with a soiled old cloth bag tied round his head. He uses the bag to capture snakes and other specimens. He's smudged, sooty, sweaty, and looking pretty much like an extra in a pirate movie. The only thing missing is the cutlass in his teeth.

But on this day he is the star of a documentary film being made about the park by the British Broadcasting Corporation. The producer is trying to keep up with Janzen, scouting locations before filming, getting to know the man known in Costa Rica as "El Professor."

Janzen is lecturing on the hoof to a group of schoolchildren, when, from the corner of his eye, he sees a squiggle of shade and light in the undergrowth. Moving over, he finds a fat, seven-foot boa constrictor. Dangling his specimen bag in front of the boa's face, he tests its aggressiveness, and when it gives only a halfhearted strike, Janzen picks up the reptile by the tail and drags it out of the vegetation as though it were a piece of old hose. He sits in the road, coils the serpent around his body, and starts petting it affectionately, telling his audience, "The amazing thing about this guy isn't that he's so dangerous, but that he's so timid and defenseless. See now, if I'd been a hungry peccary walking by . . . I often wonder how these animals survive out here."

The kids ogle the massive snake. The braver ones wrap

it around their necks like a feather boa, and take photos. Janzen starts telling a story about the boas he lets live in the rafters of his station house in Santa Rosa to get rid of the rats and toads. The BBC producer is in heaven.

"Dan," she says, "do you think it would be possible—that is, could we film you capturing the snake just as you have this morning?"

"Sure," Janzen agrees with cheerful readiness. He is not unaware of the value of good publicity to ecological campaigns. "Sure, no problem at all." He carefully coils the boa constrictor into his bag, knots the top, and stores it in the shade.

Next day the BBC film crew arrives. They set up their Arriflex on a stout wooden tripod a little way back in the forest. The producer directs Janzen to stroll down the road with another class of local schoolchildren he's showing through Santa Rosa. An assistant will release the snake so that Dan can "find" it again on camera.

Camera. Action. But this time the boa comes out of the bag with an attitude. Before Janzen can get there, the snake snaps at the assistant, then slithers off into the bush, out of range of the camera.

"Cut," says the BBC producer, as Janzen finally recaptures the snake. "Ah, Dan, do you think we could try that again?"

Janzen backs up, then saunters down the road on cue. Once again the assistant releases the boa. It heads straight for the nearest hole. When Janzen scurries after it, the snake bites him on the hand.

"Cut! Cut!—Dan, um, could we please go through that just *one* more time?"

"Sure," he says, but his eyes dart down tensely to his wristwatch: thirty minutes lost to the lifetime work plan: this is the horror of getting the world to listen.

This is how Dan Janzen plays the Guanacaste National Park game: When it is autumn in Philadelphia, he teaches at Penn and holes up in his cluttered campus office. Several times a week he may rush to the airport to fly where he is invited to lecture. He assiduously works the symposium and conference circuit. He dashes off grant proposals and palavers with the foundations. He talks on the phone with his wife, Winnie Hallwachs, a biologist who studies agoutis (a rodent related to the guinea pig) in Santa Rosa. Then he whisks down to Costa Rica for a weekend.

As soon as he gets there, he tears into the field, checks his moth traps, checks his lyomis (spiny pocket mouse) traps, checks his trees. He meets with local farmers, meets with the park rangers, meets with his "slaves" (research assistants), meets with Costa Rican government officials, solves the immediate critical problems, burns new strips around the forest, conducts an interview or three, writes a technical paper, eats a plate of rice and beans, and kisses his wife hello-goodbye.

By Monday morning he has jetted back to civilization and staggers into his classroom with ash and soot still clinging to his smoky clothes. His students are surprised he does not collapse during class. He shows them slides of butterflies, snakes, flowers, insects, trees—incredible slides that make his students his worshipers. He tells them natural history stories about the tropics. His energy can be intimidating. His scope is awesome.

At the end of the semester, everything reverses. Now he holes up with wife and the pets of the moment (from peccaries to boas) at his Santa Rosa research station. Under the corrugated metal roof, where snakes and toads and lizards and bats enter and exit at will, the Macintosh runs continuously. Thousands of moth specimens in plastic bags are strung from the raw wood rafters like tiny Japanese lanterns. Janzen is up at 4 a.m. The breakfast menu is rice and beans, with popcorn for lunch, and rice and beans for dinner. Unless, of course, there's a guest, in which case there's rice and beans for lunch, too.

Now he's on the telephone, which he installed with part of his Crafoord Prize. Several times a day he calls his office in Philadelphia, past donors and potential donors, conservation organizations, scientists he is collaborating with on various studies, and journalists, who often begin by thinking Janzen is a demented genius and sometimes end as converts to his dream.

A seemingly endless stream of visitors must be guided through Santa Rosa and initiated into the campaign, from the skinniest Costa Rican schoolchild to the deepest-pocketed American. The slide carousel continues to revolve any time Janzen is asked to lecture. He deals with local landowners, races up to San José, the Costa Rican capital, to keep in touch with national parks officials, picks up or drops off visitors or students at the airport, and sometimes flies to the States for a few days to lobby for funds, attend a conference, or receive yet another honor.

On the door of Janzen's office at Penn is a cartoon. It shows an executive behind a massive desk, ordering his secretary, "Print up 10,000 copies and send it to all the most important people in the world." But Dan Janzen is no executive sitting behind a big desk. He is the philosopher king of the jungle, the general commanding the campaign, but also the footsoldier, the horse—and the man who follows with a broom, to clean up after the horse.

CLICK.

Through the mists of time and the mysteries of Kodak, a woolly young Dan Janzen appears before our eyes. The beard and hair are thicker, blacker. The gaze is not as sure of itself, but not as weary, either. The photo is of somewhere in the jungle, and Janzen stands there with a sly grin, a revolver strapped to his hip, next to the hung carcass of a freshly shot peccary. The peccary weighs about a hundred pounds. Janzen went into the forest in the morning and returned with lunch for thirty. This was forty years ago.

"In those days," Janzen reminisced to his class at Penn, "the tropical forest was a meat market and a lumber yard. I used to chain-saw huge hardwood trees, just to count the rings for my research, with no more thought than you'd flick an insect off your sleeve."

In those days no one, including Janzen, thought it part of a tropical biologist's job to save anything. They were more concerned with herbiciding pastures for the cattle industry, or keeping bugs off the company's banana plantations. How such a big idea as growing a tropical forest took root in Dan Janzen's mind is a somewhat complex story. For one thing, he comes from an unusual background, where the ethics of hunting and conservation, scholarship and public policy, were blended.

"When I was growing up at the edge of Minneapolis, there were no suburbs," he said. "I shot my first pheasant off the back porch of our house. The city stopped where the forest began. But I could go out the front door, catch a bus, and be at the Minneapolis Public Library in fifteen minutes."

Guns and books were in his family, too. Janzen's father was director of the U.S. Fish and Wildlife Service. His parents encouraged young Dan to take to the woods with his shotgun in one hand and identification guides in his rucksack. Nor was he discouraged when he fell in love for the first time—with collecting butterflies. By the time he was in high school, Janzen was permitted to pack off alone to Veracruz, Mexico, to hunt butterflies and camp under the stars, a seminal experience for the future tropical biologist.

In the public sector of conservation, family history meant something crucial to Dan Janzen: he was steeped in a

Midwest populism that believes government is what really matters to conservation in the long run. In the early 1970s, when the government of Costa Rica began its aggressive program to establish a system of national park reserves, making that nation the leader in tropical conservation, Janzen moved his research operations to Santa Rosa National Park, in the northwest corner of the country, only a few miles from the Nicaraguan border. At Santa Rosa, Janzen began the meticulous detailing and cataloguing of the environment around him, accomplished by patiently observing, collecting, testing, and asking original questions, in the mold of the great nineteenth-century Victorian descriptive naturalists.

When Janzen came to Santa Rosa, for example, most of the three thousand or so tropical moths of the neotropical dry forests had at least been named. Some pupae and larvae had been collected. But no one had devoted the time necessary to match the adult moths with their earlier stages. For most naturalists, in itself that would be considered a lifetime's work. Night after night, year after year, Janzen has collected moths. He attracts them to simple traps made of a white sheet with a light behind it. Someone must check the sheet every hour through the night. Janzen buys the sheets in volume when Philadelphia hotels go bankrupt. He persuaded the taxonomy departments at the Smithsonian and the British Museum to pursue this great classification project. He continues to provide the specimens.

But this wasn't enough. With a whole national park to play in, Janzen began looking into another area which would become the passion of his mature years: how tropical plants and animals depend on each other for survival and reproduction. Plants and trees use insects for defense; moths, butterflies, and birds for pollination; frugivores for dispersing their seeds. Insects use plants and trees for housing. Moths and butterflies get nutrition from flower nectars. Bats, birds, and mammals concentrate on eating fruits for sustenance. These basic themes evolve in endless variety.

Janzen counted roughly 3,000 species of plants in the Santa Rosa dry forest. The same area contains 5,000 species of moths and butterflies, 300 species of birds, and about 150 types of mammals. Janzen decided to find out which animals eat which fruits, and what happens to the seeds. In the plain Midwest language he speaks, Janzen wanted to find out "who shits what where in the Neotropics."

Janzen spent much of the next twenty years down on his hands and knees, washing and sifting through the dung of peccary, horse, cattle, monkey, and mouse, looking for seeds the way miners look for gold nuggets. But while his multifarious research operations proceeded at Santa Rosa, something else was happening. Central America's tropical wildlands were rapidly disappearing to make way for cotton plantations and cattle ranching. Despite its excellent record in creating the finest national park system in the tropics, Costa Rica also has one of the highest rates of deforestation outside those park boundaries.

"We were witnessing the great change, the closing of the tropical frontier," Janzen said. "The forest was literally melting away around us. The forest I used for my own early studies in another part of Costa Rica is completely gone now. It's a sugar plantation."

An early alarm of the pressures on tropical wilderness areas naturally went off in Janzen's mind first. As prime, and then marginal, tropical lands were burned off or clear-cut for conversion, Janzen began to witness the human invasion of Santa Rosa and other national parks in the form of squatting, poaching, and slash-and-burn agriculture. The Costa Rican government could not effectively protect its own natural reserves system.

In 1985 the Costa Rican National Park Service asked Janzen to go to Corcovado National Park, in the south of the country, to report on the impact that one thousand illegal gold miners were having on the park's ecosystem. Janzen hiked through the park for ten days, interviewed almost a hundred miners, and saw firsthand the sad results of establishing a national park system without first teaching people the fundamental principles of their own environment. The gold miners were tearing up the place—cutting down trees, altering water courses, dumping sand wastes. Janzen walked for almost ten days without smelling a single peccary herd—and Dan Janzen can *smell* peccaries! Yet, almost to a man, the miners believed they were doing no harm to the environment; the majority even supported a strong national park system for Costa Rica.

From then on Janzen saw that the human vector was impossible to stop. Eventually his own study site at Santa Rosa was bound to disappear. In his introduction to the insects of Costa Rica in his book *Costa Rican Natural History*, Janzen wrote, "When I first collected moths at the lights in Guanacaste in 1963 and 1965, the gas station lights in Cañas and Liberia were a rich source of sphingids, saturniids, and many small moths. Now in the early 1980's, there is virtually nothing at these lights even at the best time of year. . . . The area of Guanacaste in original forest, heavily brushy pasture, and second-growth forest, has easily been cut by half, if not more, since 1963–65. Around the towns and other light sources the area of woody vegetation has been reduced by at least 90 percent. I suspect that rather than moths no longer being attracted to lights, there are few if any moths left."

In the dispute over the gold miners in Corcovado—the Park Service ultimately evicted the miners—Janzen saw the coming darkness if tropical wilderness must be justified on a market basis. But he saw the light as well: a management system integrated into the population, based on universal education; a park service united with ecologists, biologists, botanists, and others using the natural systems for research; restoration crews to make firebreaks and tend horse herds used to spread seeds, keep poachers out, and guide visitors; the vigorous implementation of restoration ecology action plans; the systematic collection of dung!

And he saw that tropical biologists had to act. "Within the next 10 to 30 years," Janzen wrote in a paper called "The Future of Tropical Ecology," "whatever tropical nature has not become embedded in the cultural consciousness of local and distant societies will be obliterated to make way for biological machines that produce physical goods for direct human consumption. In short, biologists are in charge of the future of tropical ecology. If the tropics of the world go under, the biologists of the world have no one but themselves to blame. . . . It is up to us to make the world conscious of its interactions with the tropical living world. If we cannot set aside our personal interests, research and development, and put our entire effort to affixing permanently some of tropical nature, then we have sold the tropics' long-term fitness for a handful of instant gratification. We are the generation for whom the only message for a tropical biologist is: *Set aside your random research and devote your life to activities that will bring the world to understand that tropical nature is an integral part of human life.*"

CLICK.

An old hacienda appears on the screen. On the veranda are some folks, some hammocks, some pet dogs. Dan Janzen is giving his slide lecture to a conservation club at DuPont's Wilmington, Delaware, plant on their lunch break. "Conservationists interested in the tropics are always asking for a laundry list of species, to decide whether a given area is worth preserving," Janzen tells them. "I don't spend much time on lists. What's the most crucial and characteristic animal of the tropical dry forest? Well, if you look at this slide you'll see them; they are human beings."

Why save tropical forests, tropical wilderness? The zone of the world lying between the Tropic of Cancer and the Tropic of Capricorn contain the planet's biological and genetic heritage in all the millions of species that humanity has previously relied on to develop food crops, medicines, and some of our most important manufactured products. As Janzen lists them, those tropical products include "chickens, eggs, elephants, turkeys, beef, pyrethrum, corn, rice, coffee, corsage orchids, tea, chocolate, morphine, strychnine, parrots, bamboo, macadamia nuts, rum, peppers, honeybees, vanilla, milk, cinnamon, dates, quinine, rubber, gardenias, bananas, avocados, mahogany, pineapples, impatiens, humans, sorghum, rosewood, coconuts, Brazil nuts, peanuts, potatoes, sweet potatoes, tapioca, squash, chimpanzees, pumpkins, beans, cane sugar, molasses, tomatoes, cats, guinea pigs, citrus, white rats, palm oil, and rhesus monkeys."

"It is a useless conceit," Janzen wrote in "The Future of Tropical Ecology," "to think our ancestors recognized more than a minute fraction of the useful products produced by the millions of species of organisms still surviving in tropical wildlands. . . . Put most simply, the tropics contain millions of species of organisms that could be managed to produce products of use to humanity." Think of tropical forests as nature's great storehouse of genetic information, belonging to all humanity, most of which has yet to be catalogued or investigated. How many potential cures for cancer are out there right now? How many plants with the potential to become the next rice, the next corn?

This is the essential argument for maintaining biodiversity in the tropics, but Janzen has another goal for tropical forest conservation. As tropical areas become developed and settled, their human inhabitants become alienated from the myriad wondrous marvels of tropical nature, the most intricate and complex ecosystems on earth. They see the consumer habits of extratropical societies, and they mimic them. Couch potatoes in paradise. Those who actually plant the cotton and cut the sugar, pick the coffee and harvest the rice lead lives Janzen describes as "frankly boring."

Janzen has learned that you cannot salvage tropical forests only for scientific and commercial reasons. You must sensitize people of the North *and* South to the great living classrooms of nature. You must tell natural history stories that add meaning to their lives. He urges those who can contribute financially to tropical conservation to go and experience tropical nature firsthand, as scientific tourists. More important, those who live nearby need to understand the processes of their neighborhood natural systems, not just see a green, irrelevant blur.

"The most important function of remaining tropical wildlands" he has written, "is to be, for very large numbers of people, the next generation's national library, theatre, church, university, and rock concert, all tied into one." The most important goal of economic development in the tropics, Janzen believes, "is not to produce a large number of barely literate human draft animals, but rather a much smaller number of intellectually well-developed humans with the ability and inclination to enjoy the riches of humanity."

CLICK.

A forbidding rock outcrop rises off the Guanacaste plains. At the top is a fortresslike ranch house with a priapic radio tower ringed in barbed wire. The place has the feel of the villain's hideaway in a James Bond movie. On the steep driveway that cuts back across the rockface, Janzen stands talking to Don Luis Gallegos, one of the landowners near the national park.

But they're not talking, they're arguing. It's not a friendly argument, either. It's about land boundaries. Janzen grabs a stick and slashes his version of the map in the dust at Gallegos's feet. Gallegos laughs and wipes it out with his foot. Gallegos is a local land baron, an honorary officer in Costa Rica's paramilitary *Guardia*. His toughs have been caught stealing horses on Guanacaste Park lands, and are suspected of setting fires to burn forest. A nasty piece of work, this Gallegos. Janzen jumps, waving his arms around. Gallegos just flicks his cigarette butt away, folds his tent, and walks back uphill. This is one good way to start a range war.

At the foot of the drive, in the parked Land Cruiser, Janzen's wife Winnie says, "I guess we're not getting invited in for Don Luis's brandy today. When we first started dealing with Gallegos, he was all sweetness and light. Used to invite us in and insist we stay for lunch. Then, after we settled the deal for his thousand-hectare ranch, Gallegos came back with a new survey and claimed the property was larger than the thousand hectares he'd sold, so he was entitled to keep the rest. 'The rest' happened to include a stream Gallegos wants to water his cattle. He may have thought he could snooker the Gringo professor, but Dan has actually learned the property laws. They're on our side."

Learning the law and customs in a country where land titles, surveys, and boundaries are more questions of personal power and political pull than legal code was also tough. It required learning how power flows in third-world countries, how to move a government bureaucracy, the ins and outs of Central American land purchases and financing. If he were not such a good ecologist, with such a highly developed sense of how parts of complex systems interact, Janzen could never have succeeded in making Guanacaste National Park a reality.

In the event, the breaks came down on Janzen's side in his crusade to restore this tropical dry forest. As he barnstormed and direct-mailed his way across donor country, the Costa Rican minister of natural resources, Alvaro Umana, informed him that the Costa Rican government would match whatever Janzen raised abroad dollar for dollar. Janzen had vowed not to take money from a poor country struggling to make its foreign debt payments—a country, he likes to point out, whose total national budget is about the same as the University of Pennsylvania's annual expenditures.

Costa Rica's foreign debt played into Janzen's hands. By the 1980s the multinational banks had so much outstanding third-world debt, considered shaky if not worthless, that they routinely discounted it and sold it back to the debtor nation just to unload it. Better to get something than nothing. This meant that with one dollar in cash, Costa Rica could go to the banks and buy back around three dollars of its debt, thereby reducing interest payments and principal. Thus for every hard dollar Janzen raised in the States, Costa Rica could retire one dollar of debt, return the original dollar to Janzen, and contribute a matching dollar to the Guanacaste Park campaign. Janzen also found that this sort of debt-for-nature swap pleased American foundations, which prefer matching grants to outright funding.

Probably the most important factor in Janzen's success, however, was that Costa Rican authorities dealing with national parks were delighted with Janzen's campaign, his action plan, and especially his restoration ecology techniques. The Park Service adopted his fire-control program at other sites to maintain habitat viability. His research and ideas on seed dispersal are also under consideration to restore other degraded ecosystems.

Mario Boza, director of the Fundacion Neotropica, praised Janzen for being "so enthusiastic, so energetic. We are also pleased that he has included $3 million for the new park's endowment, because without that, at the end, we would inherit a big problem—how to protect and maintain that large area without the proper personnel, equipment, facilities and procedures. Janzen has become the godfather of Guanacaste National Park. There are many scientists doing research in our parks who are exclusively interested in gathering the data they need, and forget about the park. We would like to have other Dan Janzens as godfathers for our other national parks."

It might have been otherwise—a knee-jerk nationalistic response: Where does this Gringo academic with the headrag and the *eau de peccary* odor come off telling us what to do with our national patrimony? The head-butting session with the landowner showed how that might have gone. And despite the fact that Guanacaste National Park became a reality, Janzen still must scramble. Time does not stand still. The population bomb is ticking. Agribusiness could pressure the government. One of the volcanoes might erupt. Or precious metals might be discovered under the park. As Janzen states religiously at every opportunity, it is simply not enough in the American tropics to raise funds, put a fence around the forest, and call it preserved. Economic and social pressures will inevitably unpreserve it.

"The challenge isn't what is going to be preserved of tropical forests now," said Janzen. "Those decisions have already been made, that game's over. The challenge is what will be left of tropical forests a hundred years from now. And I don't care how much money the Nature Conservancy or the World Wildlife Fund puts into it, if the people don't understand it and want it, there won't be any national parks a hundred years from now."

CLICK.

At Santa Rosa in Costa Rica, while stealing an hour to measure the tree growth in a stand of fifty-year-old hymeneas he has been tracking, Janzen was asked if he knows every tree in Santa Rosa's dry forest.

"You have to remember," he replied, "I've put the same energy into this forest that other people put into their kids, their careers, their homes. I'm sure a housewife in Philadelphia knows all the different little shops to find certain things in her neighborhood. Well, this forest is my home, my neighborhood, my family, my library."

We walk back through his rec room—a riverbed, dry as bone. A tinamou scoots away from underfoot. White-faced monkeys come down through the trees to watch. Janzen entertains himself by picking small fruits off the ground. He pops them into his mouth, swallows the seeds, and spits out the pulp—the opposite of what the rest of the animal kingdom does. Another experiment, or just Janzen being Janzen? Hard to tell.

Acknowledgments

THIS BOOK, which is about tropical nature, tropical forests, tropical rivers, and tropical reefs, ironically owes its existence mostly to libraries and librarians, who have helped me obtain access to accounts of tropical explorations long out of print. Without these guides I could not have completed the book. The collection of this raw material, though only enough to fill a few bookshelves, actually took more than five years to accomplish. It was seeded by the John Simon Guggenheim Memorial Foundation, to whom I wish to express my chief indebtedness. The original idea for the book hatched in stimulating conversations with my stalwart friend, the writer Doug Hand, to whom I am more than grateful.

My reading itinerary began at the University of South Alabama Library in Mobile, where Professor Lloyd Dendinger and Dr. Bob Shipp introduced me to Archie Carr's books and the biology of sea turtles. Larry Ogren of National Marine Fisheries Services, Carr's student and a sea turtle man himself, was kind enough to take the time to reminisce about his former professor.

At the Watson Library at Northwestern State University in Natchitoches, Louisiana, rare book librarian Mary Lynn Banderas loaned me a second English edition of Humboldt's *Narratives of Travel* in six volumes. I spent two subtropical winters there reading it, supported by my friend

Grady Ballenger, whom I thank for feeding and paying me, going into the bayous in canoes with me, and bringing the course "Literature of American Tropical Biology" to the Scholars College. Librarian Jan Samet was also very helpful while she was still at NSU.

The trail led next to the Latin America collection at the Howard-Tilton Memorial Library of Tulane University in New Orleans, where Guillermo Nanez in the Rare Book Room allowed me to spend time with Frederic Catherwood's gorgeous color lithographs of the pre-Columbian ruins he had discovered with John Stephens. New Orleans Public Library librarian Linda Hill also deserves thanks for pointing me in the right direction at several junctures. I am grateful, too, to Charlotte and Jean Seidenberg in New Orleans, who first introduced me to Margaret Mee's botanical illustrations of Amazonia. And to artist-environmentalist Jacqueline Bishop, who filled me in on this remarkable woman and helped me find Mee's book. My heartfelt thanks also for the Southern hospitality and scholarly guidance of Tulane botanical archivist Anne Bradburn, who fed me body and soul while I pitched my camp in New Orleans. It was through Anne that I met my friend Professor Victor Fet, to whom I am grateful for inviting me to lecture at Loyola University.

Closer to home, three libraries and their able staffs have assisted me. First thanks go to Cape May County Library, where my favorite reference librarian Marie Jones has kept the home fires burning and looked up a thousand things for me during the course of this project. My gratitude also to librarian Janet Baldwin at the Explorers Club Headquarters in New York for help in accessing William Beebe's books. Finally, thanks to the library at Richard Stockton College of New Jersey for help with maps and for keeping my part-time faculty card good year-round.

I wish to express my gratitude to the BBC library at

the Natural History Unit in Bristol, England, where Sheila Fullom connected me to a pirate's treasure of books on tropical natural history. Bless Sheila, too, for filling my pockets with five-pound notes to buy natural history books on my lunch hour in the bountiful bookstores of Bristol.

I am indebted to two editors in the writing of this book. One is Bill Tonelli, who commissioned me to write on Dan Janzen for *Philadelphia* magazine, allowing me to interview Janzen at length in Costa Rica and collect and read a dozen of his books and papers. The other is Noel Young of Capra Press in Santa Barbara, for commissioning me to write on W. H. Hudson, thus enabling me to afford to buy his books in old and reprint editions.

Much love to my dear cousin Leonard Baskin for long ago giving me W. H. Hudson to read, which influenced me to study natural history, and for supporting this book from its earliest conception.

Thanks, finally, to Arseniy Khobotkov for his bibliographical and indexing efforts.

Selected Bibliography

Isaac Asimov, *A Short History of Biology*, New York: Natural History Press, 1964.
Edward Ayensu, ed., *Jungles*, New York: Crown, 1980.
Michelle Barr, Rick Bergeron, Amy Pelt, and Peter Rolufs, *Measuring America: A Selection of Alexander von Humboldt's Scientific Instruments*, Natchitoches, La.: Anabasis Press, 1992.
William Beebe, *The Arcturus Adventure*, New York: Knickerbocker Press, 1926.
William Beebe, *Beneath Tropic Seas*, New York: G. P. Putnam's Sons, 1927.
William Beebe, *Half Mile Down*, New York: Harcourt, Brace and Company, 1935.
David Bellamy, *Bellamy's New World*, London: British Broadcasting Corporation, 1983.
Richard Bitterling, *Alexander von Humboldt*, Berlin: Deutscher Kunstverlag, 1959.
Harriet Bridgeman and Elizabeth Drury, *The British Eccentric*, New York: Clarkson N. Potter, 1975.
Archie Carr, *The Sea Turtle*, Austin: University of Texas Press, 1984.
Archie Carr, *So Excellent a Fishe: A Natural History of Sea Turtles*, New York: Anchor Press, 1973.
Archie Carr, *The Windward Road*, New York: Alfred A. Knopf, 1963.
Ronald W. Cox, *Explorers of the Deep*, Maplewood, N.J.: Hammond, 1968.
Kathleen Curl, Natasha Munsen, and Jennifer Porche, *A Waterton Bestiary*, Natchitoches, La.: Anabasis Press, 1993.
Charles Darwin, *On the Origin of Species*, Cambridge, Mass.: Harvard University Press, 1964.
Charles Darwin, *The Voyage of the Beagle*, New York: Penguin, 1989.

Stephen Jay Gould, "Church, Humboldt, and Darwin: The Tension and Harmony in Art and Science," in "Fredric Edwin Church," catalog of an exhibition, Washington, D.C.: Smithsonian Institution Press, no date.

Stephen Jay Gould, *Ever Since Darwin*, New York: W. W. Norton, 1979.

Victor Wolfgang von Hagen, *Maya Explorer*, San Francisco: Chronicle Books, 1983.

Dorothy and Bob Hargreaves, *Tropical Blossoms of the Caribbean*, Kailua, Hawaii: Hargreaves, 1960.

Dorothy and Bob Hargreaves, *Tropical Trees*, Kailua, Hawaii: Hargreaves, 1965.

W. H. Hudson, *Far Away and Long Ago*, New York: Hippocrene Books, 1984, facsimile edition.

W. H. Hudson, *Idle Days in Patagonia*, Berkeley, Calif.: Creative Arts, 1979.

W. H. Hudson, *The Naturalist in La Plata*, New York: AMS Press, 1968.

Alexander von Humboldt, *Aspects of Nature*, New York: AMS Press, 1970.

Alexander von Humboldt, *Personal Narrative of Travels to the Equinoctial Regions of the New Continent During the Years 1799–1804*, trans. Helen Maria Williams, London: Longman, Hurst, Rees, Orme, and Brown, 1818.

Alexander von Humboldt, *Political Essay on the Kingdom of New Spain*, Norman: University of Oklahoma Press, 1988.

Thomas H. Huxley, *American Addresses*, New York: D. Appleton, 1893.

Daniel H. Janzen, *Costa Rican Natural History*, Chicago: University of Chicago Press, 1983.

Daniel H. Janzen, *Ecology of Plants in the Tropics*, London: Butler and Tanner, 1975.

Daniel H. Janzen, *Guanacaste National Park: Tropical Ecological and Cultural Restoration*, San Jose, Costa Rica: Editorial Universidad Estatal a Distancia, 1986.

Daniel Janzen and Paul S. Martin, "Neotropical Anachronisms: The Fruits the Gomphotheres Ate," *Science*, vol. 215, January 1, 1982, 19–27.

Alan C. Jenkins, *The Naturalists*, New York: Mayflower Books, 1978.

Larousse Encyclopedia of the Earth, Buffalo: Prometheus Books, 1961, pp. 110–112.

Arthur Maurice, "The Birds' Best Human Friend—W. H. Hudson," *Natural History*, 1924.

David McCullough, "The Man Who Rediscovered America," *Audubon*, September 1973, pp. 51–63.

Beth McGeachy, *Handbook of Florida Palms*, St. Petersburg, Fla.: Great Outdoors, 1982.

Marianne North, *Recollections of a Happy Life*, New York: Macmillan, 1893.

Donald Perry, *Life Above the Jungle Floor*, New York: Simon and Schuster, 1986.

Frances Perry and Roy Hay, *A Field Guide to Tropical and Subtropical Plants*, New York: Van Nostrand Reinhold, 1982.

Fritz Sandy, "The Living Reef," *Popular Science*, May 1995, pp. 48–51.

Helen and Frank Schreider, *Exploring the Amazon*, Washington, D.C.: National Geographic Society, 1970.

Edith Sitwell, *English Eccentrics*, New York: Vanguard, 1957.

Alexander F. Skutch, "The Advancement of Biology in the Tropics," Bulletin No. 4, New York Botanical Garden: Association for Tropical Biology, 1965.

Alexander F. Skutch, *Birds of Tropical America*, Austin: University of Texas Press, 1983.

Alexander F. Skutch, *Helpers at Birds' Nests*, Iowa City: University of Iowa Press, 1987.

Alexander F. Skutch, *The Imperative Call*, Gainesville: University Presses of Florida, 1979.

Alexander F. Skutch, *A Naturalist on a Tropical Farm*, Berkeley: University of California Press, 1980.

Alexander F. Skutch, *Nature Through Tropical Windows*, Berkeley: University of California Press, 1983.

Jeremy Stafford-Deitsch, *Reef: A Safari Through the Coral World*, San Francisco: Sierra Club Books, 1991.

John L. Stephens, *Incidents of Travel in Central America, Chiapas and Yucatan*, New York: Dover, 1969.

John L. Stephens, *Incidents of Travel in Yucatan*, New York: Dover, 1963.

William Stolzenburg, "The Old Man and the Sea," *Nature Conservancy*, November/December 1994, pp. 16–23.

Kenhelm W. Stott, Jr., "Profiles from the Past," *Explorers Club Newsletter*, 1981, vol. 8, no. 1.

Gordon J. Veath, "It Broke the Depth Barrier," *Cue*, April 26, 1969, pp. 11–12.

Alfred Russel Wallace, *A Narrative of Travels on the Amazon and Rio Negro*, London: Ward, Lock, Bowden, 1892.

Charles Waterton, *Wanderings in South America*, New York: Macmillian, 1893.

Charles Waterton, *Wanderings in South America*, New York: Hippocrene Books, 1983.

Index

A NOTE ON THE AUTHOR

Writer, naturalist, and filmmaker, Jonathan Maslow has also written *The Owl Papers*; *Sacred Horses: Memoirs of a Turkmen Cowboy*; and *Torrid Zone*, a collection of stories. His book *Bird of Life, Bird of Death* was nominated for a National Book Critics Circle nonfiction prize and received the Harold D. Vursell Memorial Award from the American Academy of Arts and Letters. Mr. Maslow was born in New Jersey and studied at Marlboro College and the Columbia Graduate School of Journalism. He has also been a Guggenheim Fellow. He teaches regularly at several colleges, travels widely, and writes frequently for such publications as the *Atlantic Monthly*, the *New York Times*, *Life*, *Reader's Digest*, *Country Journal*, and *BBC Wildlife*. He lives in Cape May, New Jersey.